Frontier Assemblages

Antipode Book Series

Series Editors: Vinay Gidwani, University of Minnesota, USA and Sharad Chari, University of California, Berkeley, USA

Like its parent journal, the Antipode Book Series reflects distinctive new developments in radical geography. It publishes books in a variety of formats – from reference books to works of broad explication to titles that develop and extend the scholarly research base – but the commitment is always the same: to contribute to the praxis of a new and more just society.

Published

Frontier Assemblages

The Emergent Politics of Resource Frontiers in Asia

Edited by

Jason Cons

Michael Eilenberg

WILEY

Registered Office(s)
John Wiley & Sons, Inc., 111 River Street, Hoboken, NJ 07030, USA
John Wiley & Sons Ltd, The Atrium, Southern Gate, Chichester, West Sussex, PO19 8SQ, UK

Editorial Office
9600 Garsington Road, Oxford, OX4 2DQ, UK

For details of our global editorial offices, customer services, and more information about Wiley products visit us at www.wiley.com.

Wiley also publishes its books in a variety of electronic formats and by print-on-demand. Some content that appears in standard print versions of this book may not be available in other formats.

Library of Congress Cataloging-in-Publication data applied for

ISBN: 9781119412069 (hardback)
ISBN: 9781119412052 (paperback)

Cover image: Frontier infrastructure, newly asphalted road crossing a bridge into forest interior of West Kalimantan, Indonesia © Michael Eilenberg
Cover design by Wiley

Set in 10.5/12.5pt Sabon by SPi Global, Pondicherry, India
Printed in Singapore by C.O.S. Printers Pte Ltd

10   9   8   7   6   5   4   3   2   1

Contents

List of Figures

Series Editors' Preface

The *Antipode Book Series* explores radical geography 'antipodally', in opposition, from various margins, limits, or borderlands.

Antipode books provide insight 'from elsewhere', across boundaries rarely transgressed, with internationalist ambition and located insight; they diagnose grounded critique emerging from particular contradictory social relations in order to sharpen the stakes and broaden public awareness. An *Antipode* book might revise scholarly debates by pushing at disciplinary boundaries, or by showing what happens to a problem as it moves or changes. It might investigate entanglements of power and struggle in particular sites, but with lessons that travel with surprising echoes elsewhere.

Antipode books will be theoretically bold and empirically rich, written in lively, accessible prose that does not sacrifice clarity at the altar of sophistication. We seek books from within and beyond the discipline of geography that deploy geographical critique in order to understand and transform our fractured world.

Vinay Gidwani
University of Minnesota, USA

Sharad Chari
University of California, Berkeley, USA

Antipode Book Series Editors

Notes on Contributors

Zachary R. Anderson is a PhD candidate in the Department of Geography at the University of Toronto. He has conducted research on the cultural politics of conservation, development, and resource extraction in frontier spaces across Southeast Asia. His doctoral research investigates the emergence of the 'green economy' in Indonesia in the province of East Kalimantan. He has published in *Journal of Peasant Studies,* the *Austrian Journal of South-East Asian Studies, Global Environmental Change,* and *Conservation Biology.*

Young Rae Choi is an assistant professor of Geography in the Department of Global and Sociocultural Studies at Florida International University. Her research interrogates the complexity and interwovenness of development-conservation relations with a focus on large-scale coastal development in East Asia. Previously, she worked on marine policy and strategic planning of ocean science research at the Korea Institute of Ocean Science and Technology and led the Korean side of the Worldwide Fund for Nature Yellow Sea Ecoregion Support Project as the national conservation coordinator. Her work has been published in *Ocean & Coastal Management, Dialogues in Human Geography,* and *Marine Pollution Bulletin.*

Jason Cons is an assistant professor of Anthropology at the University of Texas at Austin. He works on borders in South Asia, especially the India–Bangladesh border; on agrarian change and rural development in Bangladesh; and, most recently on climate change, development, conservation, and security along the India–Bangladesh border. His book, *Sensitive Space: Anxious Territory at the India–Bangladesh Border* was published by the University of Washington Press in 2016. His work has also been published in *Antipode, Cultural Anthropology, Ethnography,*

Journal of Peasant Studies, Limn, Modern Asian Studies, Political Geography, SAMAJ, and *Third World Quarterly.* He is an Associate editor of the journal *South Asia.*

Prasenjit Duara is the Oscar Tang Chair of East Asian Studies at Duke University. In 1988, he published *Culture, Power and the State: Rural North China, 1900–1942* (Stanford University Press) which won the Fairbank Prize of the AHA and the Levenson Prize of the AAS, USA. Among his other books are *Rescuing History from the Nation* (University of Chicago Press, 1995), *Sovereignty and Authenticity: Manchukuo and the East Asian Modern* (Rowman 2003), and most recently, *The Crisis of Global Modernity: Asian Traditions and a Sustainable Future* (Cambridge University Press, 2014).

Michael Eilenberg is an associate professor of Anthropology at Aarhus University. His research focuses on issues of state formation, sovereignty, autonomy, citizenship and agrarian expansion in frontier regions of Southeast Asia with a special focus on Indonesia and Malaysia. His book, *At the Edges of States,* first published by KITLV Press (2012) and later reprinted by Brill Academic Publishers (2014), deals with the dynamics of state formation and resource struggle in the Indonesian borderlands. His articles have appeared in *Asia Pacific Viewpoint, Identities: Global Studies in Culture and Power, Journal of Borderland Studies, Journal of Peasant Studies, Modern Asian Studies* and *Development and Change.*

Gökçe Günel is an assistant professor in the School of Middle Eastern and North African Studies at the University of Arizona, and specializes in social studies of energy and climate change. She is the author of *Spaceship in the Desert: Energy, Climate Change and Urban Design in Abu Dhabi* (Duke University Press, 2019). Her articles have appeared in *Ephemera, Public Culture, Anthropological Quarterly, The Yearbook of Comparative Literature, The ARPA Journal, Avery Review, International Journal of Middle Eastern Studies (IJMES), Engineering Studies,* and *PoLAR.*

Christian C. Lentz is assistant professor of Geography at the University of North Carolina at Chapel Hill. He specializes in Southeast Asia with particular focus on agrarian studies, development, state formation, nationalism, and nature-society relations. His articles have appeared in *Geopolitics, Journal of Vietnamese Studies, Political Geography, Modern Asian Studies,* and *Journal of Peasant Studies.* His book manuscript *Contested Territory: Dien Bien Phu and the Making of Northwest*

Vietnam, forthcoming with Yale University Press (2019), explores hidden histories of territorial construction and political struggle during and after the battle that toppled French Indochina in 1954.

Christian Lund is a professor of Development, Resource Management, and Governance, at the Department of Food and Resource Economics, University of Copenhagen. His research focuses on property, local politics and state formation; in particular socio-legal processes of conflict over land and natural resources. He is the author of *Law, Power and Politics in Niger: Land Struggles and the Rural Code* (Lit Verlag/ Transaction Publishers) and *Local Politics and the Dynamics of Property in Africa* (Cambridge University Press). He currently is working on a book manuscript, *Nine-Tenths of the Law: Enduring Dispossession in Indonesia*.

Duncan McDuie-Ra is professor of Development Studies at University of New South Wales, Sydney. His most recent books include *Northeast Migrants in Delhi: Race, Refuge and Retail* (Amsterdam University Press, 2012), *Debating Race in Contemporary India* (Palgrave Macmillan, 2015) and *Borderland City in New India: Frontier to Gateway* (Amsterdam University Press, 2016). His articles have appeared in *South Asia: Journal of South Asian Studies, Geoforum, Urban Studies, Geographical Journal, Energy Policy, Men and Masculinities,* and *Violence Against Women* among others. He is associate editor for the journal *South Asia*, for the book series *Asian Borderlands* (Amsterdam University Press) and editor in chief of the *ASAA South Asia* monograph series (Routledge).

Townsend Middleton is an associate professor of anthropology at the University of North Carolina at Chapel Hill. He is the author of *The Demands of Recognition: State Anthropology and Ethnopolitics in Darjeeling* (Stanford University Press, 2015); and author of various articles in journals such as *Public Culture* (2018), *American Anthropologist* (2013), *American Ethnologist* (2011), *Ethnography* (2014), *Political Geography* (2013), and *Focaal* (2013). In addition to his ongoing research on cinchona, he is currently leading a collaborative interdisciplinary project examining logistical and infrastructural 'chokepoints' around the world and writing on topics ranging from colonial history to contemporary political violence in South Asia.

Kasia Paprocki is an assistant professor of Environment in the Department of Geography and Environment at the London School of Economics and Political Science. Her research is focused on the political ecology of development and climate change adaptation, particularly in Bangladesh. Her work has been published in academic and popular outlets including *Annals of the Association of American Geographers, Geoforum, Climate and Development, Journal of Peasant Studies, Third World Quarterly, Economic and Political Weekly,* and *Himal Southasian.*

Nancy Lee Peluso is Henry J. Vaux Distinguished Professor of Forest Policy and professor of Society and Environment in the Department of Environmental Science, Policy, and Management (ESPM) at University of California, Berkeley. Her work explores agrarian and forest politics, focusing in particular on the political ecologies of resource access, use, and control. She is the author of *Rich Forests, Poor People: Resource Control and Resistance in* Java (UC Press, 1992); and co-editor of six books, including *Violent Environments* (Cornell Press, 2001, with Michael Watts), *New Frontiers of Land Control* (2011, Routledge, with Christian Lund) and author or co-author of more than 70 journal articles and book chapters. She is currently working on a book examining historical entanglements of violence and territorialities in resource landscapes of West Kalimantan, Indonesia.

Igor Rubinov is a PhD candidate in Anthropology at Princeton University. He has conducted research on development, migration and the environment in Central Asia. His dissertation project, conducted over 16 months, examines the impact of climate change adaptation on governance and livelihoods in Tajikistan. As state and international agencies worked to incorporate this novel paradigm, people improvised material and social entanglements to nourish life. He has published in *Anthropological Quarterly.*

K. Sivaramakrishnan is Dinakar Singh Professor of Anthropology, professor of Forestry and Environmental Studies, and co-director of the Program in Agrarian Studies at Yale University. His current research includes work on environmental jurisprudence in India and urban ecology in Asia. His published work covers environmental history and political anthropology, science and technology studies, and cultural geography. He is the author of *Modern Forests* (Stanford University Press, 1999). Most recently he is the co-editor of *Places of Nature in Ecologies of Urbanism* (Hong Kong University Press, 2017).

Heather Anne Swanson is an associate professor of Anthropology at Aarhus University, deputy director of its Centre for Environmental Humanities, and a member of the Aarhus University Research on the Anthropocene project. Her work investigates entangled human and nonhuman lives in times of anthropogenic disturbance and environmental damage. She is co-editor of *Arts of Living on a Damaged Planet* (2017, University of Minnesota Press) with Anna Tsing, Elaine Gan, and Nils Bubandt. Most recently she is the co-editor of *Domestication Gone Wild* (2018, Duke University Press) with Marianne Lien and Gro Ween. She has published in *Social Analysis, Science as Culture, Environmental Humanities, Geoforum, Environment and Society,* and *HAU Journal of Ethnographic Theory.*

Max D. Woodworth is an assistant professor in the Department of Geography, Ohio State University. His research to date has focused on urban development in mining regions, with an emphasis on the politics of large-scale development projects in resource boomtowns. He has published in *The Journal of Asian Studies, The Professional Geographer, Geoforum, Cities,* and *Area.*

Jerry Zee is an assistant professor of Anthropology at University of California, Santa Cruz. His work explores experiments in environment and politics along the trajectory of dust storms in and beyond China. His work has appeared in *Cultural Anthropology, American Anthropologist,* and *Scapegoat.*

Acknowledgements

The editors would like to thank all the authors for their enthusiasm, active engagement with, and support of this project. We would like, especially, to thank the authors of the commentaries – Christian Lund, Nancy Lee Peluso, K. Sivaramakrishnan, and Prasenjit Duara—for their time and thoughtful and generous readings of these chapters. The majority of the contributors to this project first met in late April 2016 as part of the Frontier Assemblages workshop (convened by Cons and Eilenberg) at the Social Science Research Council's InterAsian Connections V conference in Seoul. This workshop served as a launching point for this volume and a generative moment for many of the ideas expressed in it. We would like to thank the SSRC's InterAsia Program – especially Seteney Shami, Holly Danzeisen, and Mona Saghri – for providing intellectual guidance and financial support, and the Seoul National University Asia Center for their hospitality in hosting us and this event. Their support made it possible to assemble such a vibrant group of scholars working on frontiers across Asia. Prasenjit Duara, Sahana Ghosh, Angela Ki-che Leung, and K. Sivaramakrishnan provided valuable questions and commentary during the workshop. Of the original participants, Mike Dwyer, Dolly Kikon, and Jasnea Sarma have not included essays in this volume. We thank them for participating in the workshop and providing support and encouragement to the project. A longer and substantially different version of Chapter 1 was published in *Antipode* Volume 51, Issue 1.

We would like to thank Lloyd Farley for his work in assembling the bibliography for this book. We received supportive and constructive feedback on the introduction from Emmy Dawson, Sabrina Lilleby Duncan McDuie-Ra, Townsend Middleton, Daniel Ng, Kenza Yousfi, and Stephen Zigmund. Erin Lentz, Vasilina Orlova, and James Slotta, provided feedback on earlier iterations of this project. We received invaluable feedback on the manuscript as a whole from two anonymous reviewers. Finally, we wish to thank Jacqueline Scott at Wiley and Vinay Gidwani and Sharad Chari of the Antipode Book Series for their support of this project and their work in bringing it to publication.

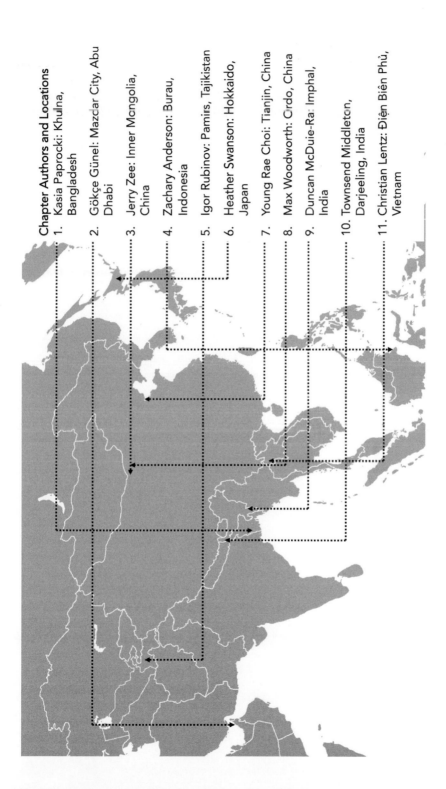

Chapter Authors and Locations
1. Kasia Paprocki: Khulna, Bangladesh
2. Gökçe Günel: Mazdar City, Abu Dhabi
3. Jerry Zee: Inner Mongolia, China
4. Zachary Anderson: Burau, Indonesia
5. Igor Rubinov: Pamirs, Tajikistan
6. Heather Swanson: Hokkaido, Japan
7. Young Rae Choi: Tianjin, China
8. Max Woodworth: Ordo, China
9. Duncan McDuie-Ra: Imphal, India
10. Townsend Middleton, Darjeeling, India
11. Christian Lentz: Điện Biên Phủ, Vietnam

Introduction

On the New Politics of Margins in Asia
Mapping Frontier Assemblages

Jason Cons and Michael Eilenberg

How shall the inhabitants of a 'remote area' evaluate the arbitrary love-hate of its visitors? Are alternative periods of 'unspoiledness' and violence their inevitable fate? – Edwin Ardener

This volume responds to the emergence, and we argue convergence, of two phenomena across Asia over the past handful of decades. The first is the rapid transformation of forest and agrarian spaces into sites of export-oriented resource extraction. Whether in the conversion of vast swaths of rainforest to oil palm and rubber plantations across Southeast Asia or the explosion of large-scale and wildcat mining operations around the Pacific rim, millions of acres of land have been rapidly converted into sites for often ecologically and socially destructive extraction.[1] The causes of this expansion are various, but broadly they have been stimulated by the search for new investment opportunities by transnational companies, both beyond but especially within Asia, and a boom in transnational investments and development collaborations anchored in global supply chains (Hall, 2011; Borras and Franco, 2011; Buchanan et al., 2013; Baird, 2014; Li, 2014b; Kelly and Peluso, 2015; Li, 2015). Alongside this unprecedented expansion have been a myriad of other transformations of remote space into new kinds of productive sites – sites slated for massive infrastructural projects, export processing zones, new urban developments, spaces of privatized health care, habitats of ecological reclamation and sustainability, speculative locations for carbon storage and more. The proposition of this volume is that these two

Frontier Assemblages: The Emergent Politics of Resource Frontiers in Asia,
First Edition. Edited by Jason Cons and Michael Eilenberg.

processes of extraction and production should be understood together as linked projects of incorporating margins and remote areas into new territorial formations. In other words, these out-of-the way places (Tsing, 1993) are key sites in the making of, and thus key vantage points for understanding, new articulations of territorial rule, regional and global networks of accumulation, and security.[2] We argue that both these productive and extractive transformations should be understood as the making of new Asian resource frontiers.

Studies of resource frontiers have primarily explored extractive spaces – areas where monocultural crop booms or the discovery of new mineral or petrochemical resources have rapidly reconfigured land tenure and sociality alongside of political economy and ecology (Sturgeon, 2005; McCarthy, 2006; De Koninck et al., 2011; Hall et al., 2011). In this volume, we move away from an exclusive focus on extraction, and understand resource frontiers also as sites of creative, if often ruinous, production. In doing so, we offer two rejoinders to the more well-trodden literature on the political economy of extraction. First, we suggest that what matters in the incorporation (or re-incorporation) of margins are the various forces and processes that are assembled to reinvent these spaces as zones of opportunity. And second, we suggest that not only are these forces of spatial transformation resonant across sites, resources, and interventions, but that a broader view of territorial intervention gives us tools to understand a moment in which the relationship of millions of people to land and rule is being radically reconfigured. Moreover, we suggest that at once plumbing the unique histories of individual frontiers and understanding similarities across different frontiers might open new possibilities for responding to exploitation.

How might we understand the forces that precipitate these sweeping transformations throughout the region? And what do these shifts portend for Asia's margins, many of which have and continue to be sites of intense securitization, instability, conflict, and expansion? What similarities and differences do these transformations share? This volume ventures a series of initial studies of these questions. Each chapter offers a rich ethnographic and/or historical study of a particular resource frontier. Yet collectively, we begin to trace broader patterns of contemporary frontier making and their effects.

To do this, we turn our attention to what we call *frontier assemblages*: the intertwined materialities, actors, cultural logics, spatial dynamics, ecologies, and political economic processes that produce particular places as resource frontiers. Frontier assemblage is a term that is both descriptive and analytic. Contributors to this volume use it to map the histories and geographies that coalesce in specific places and moments to produce resource frontiers. At the same time, we use it to raise questions about the continuities and disjunctures of what we understand as the current round of incorporating margins across Asia. Resource frontiers are sites in which new forms of territorial power are formed through the convergence of a variety of forces. They are also windows onto broader processes of managing risk, facilitating accumulation, and reconfiguring sovereignty. Through the analytic of frontier assemblage, contributors offer a perspective on such transformations that does not – a priori – privilege specific causal understandings, but augers a mapping of flows, frictions, interests, and imaginations that accumulate in particular places to transformative effect.

Asian Margins in Flux?

As a rich literature shows, the dynamic tension between centres and margins is a key trope in Asian history. Whether for purposes of settling and managing questionable populations, instituting sedentary agrarian regimes, opening up new spaces for trade and capital expansion, shoring up colonial and national security, or producing 'buffer' zones between competing empires, the production and management of margins as frontiers has been a constant and unfolding challenge in the making of Asian sovereignties, territories, and regimes of rule. The tensions of incorporating fugitive landscapes in pre-colonial Southeast Asia (von Schendel, 2002; Tagliacozzo, 2005; Scott, 2009); the imperial management of peripheries in early Modern China (Crossley et al., 2006; Bryson, 2016); the colonial attempts to settle unruly frontiers in South Asia (Bayly, 2000; Ludden, 2011; Zou and Kumar, 2011); and the politics of managing postcolonial and Cold War rivalries in upland and remote spaces throughout the continent (McGranahan, 2010; Eilenberg, 2011, 2012; Guyot-Réchard, 2016); are but a few well known moments in which marginal space has become central to regional and geo-politics. A constant throughout this frontier history has been the uncertainties, anxieties, and failures inherent in attempts to incorporate marginal spaces into logics of territorial rule. Read broadly, frontiers in Asian history emerge not just as the bleeding edge of territorial expansions and empires, but as ambiguous sites where opportunity and possibility are intimately linked to resistance and official unease. The dynamics under examination in this volume, then, might be thought of as only the current round of a much longer historical dynamic.

Yet, the scope of this current moment of frontier expansion – alongside its human and ecological costs – demands a critical interrogation of the resonances and disjunctures in the making of new resource frontiers across the continent. There are a range of proximate drivers of this

current expansion. Ongoing waves of neoliberal reform have contributed to the opening up of both economies and particular spaces to
foreign direct investment and corporate management. This is particularly apparent in the explosion of export processing and concession
zones that have emerged across Southern Asia during and in the wake of
structural adjustment programs in the 1980s and 1990s. Alongside these
dynamics of liberalization, neo-Malthusian narratives about scarcity
have heralded massive expansions of plantation-based monocultures in
marginal and upland space. A parallel Malthusian logic of energy security has wrought similar expansion. Marginal spaces across the continent are increasingly the sites of prospecting for oil, natural gas, and
coal, as well as the locations for new and often massive hydro-electric
projects. These processes constitute an important part of the broad and
much debated 'global land grab' (Dwyer, 2013; Wolford et al., 2013b;
Baird, 2014).

Collectively, these transformations in land and the meanings of both
frontiers and marginal space might be thought of as a critical conjuncture in the longer trajectory of capital in Asia and beyond. Indeed, these
phenomena have been a focal point of both activism and concerned
scholarship in Asia over the past two decades. Scholars have critically
examined the processes of producing marginal spaces as frontiers
through analyses of enclosure, concessions, and special economic zones;
the mapping of networks of national and transnational capital; and the
exploration of the networks and circuits of labour involved in resource
extraction (Tsing, 2005; Bach, 2011; Levien, 2011; Arnold, 2012;
Eilenberg, 2012). Such analyses often figure resource frontiers as spatio-
temporal fixes (Harvey, 2001): locales bound up in both producing value
and solving a range of crises of over-accumulation. In other words, the
political economy of these new resource frontiers situates them as key
sites of capital: securing its expansion and insuring against its collapse.

Yet, as contributors to this volume demonstrate, capital is only one, if
a central, force producing contemporary resource frontiers. Indeed, the
chapters to come demonstrate a range of forces, actors, and processes
equally crucial to the understanding of the current conjuncture. Many
contributors highlight the various ways that futures and presents of
environmental collapse lurk at the heart of frontier projects (Zee,
Anderson, Paprocki, Choi). Others highlight the ways that the dynamics
of imagination and fantasy shape new frontiers (Günel, Woodworth,
McDuie-Ra). Still others trace the ways that the pasts of frontier production linger on and generate new possibilities and challenges within frontier space (Middleton, Lentz, Rubinov, Swanson). Key to all of these
investigations are the ways that political economies of frontiers are
always entangled with a broader array of factors that structure the

transformation of marginal space into frontier zone. Indeed, these entanglements themselves prove to be fruitful in understanding not only the dynamics of contemporary Asian resource frontiers, but the ways these spaces do and do not articulate with each other. Mapping these dynamics, then, offers ways to not only rethink resource frontiers, but to reimagine debates over globalization, with their often-narrow focus on urban space and networks of capital circulation. Indeed, such an outlook allows us to rethink and decentre the broad geopolitical paradigms that shape existing debates over resource frontiers and to open new questions about the structures and workings of both frontier space and global flow. To better understand these dynamics, we turn to our analytic of frontier assemblages.

Assemblages and Frontiers

'Frontier assemblage' brings together two highly, some might say hopelessly, overdetermined concepts in a single phrase. Both of these terms have been explored and debated in exhaustive detail elsewhere (Prescott, 1987; Donnan and Wilson, 1994; Baud and van Schendel, 1997; Wendl and Rösler, 1999; Geiger, 2008; Nail, 2017). Rather than rehearse these debates in full, we offer a thumbnail sketch of their genealogies before making a case for understanding resource frontiers *as* frontier assemblages.

The notion of assemblage springs from the work of Deleuze and Guattari (1986, 1987). It articulates an approach to understanding compositions of various sorts (social, ecological, territorial, etc.) beyond an analysis that reduces them to simple consequences of human behaviour. The concept is notoriously open-ended. As Deleuze argues, 'an assemblage is first and foremost what keeps very heterogeneous elements together: e.g. a sound, a gesture, a position, etc. both natural and artificial elements. The problem is one of "consistency" or "coherence", and it is prior to the problem of behavior' (Deleuze, 2007: 179). Assemblage, then, is a loose theoretical framework that seeks to destabilize classical models of social theory with their emphasis on human causality, and to replace it with what Deleuze calls 'hodgepodges': contingent collections of things whose coming together itself is not the precondition, but rather the object of inquiry.[3]

Our use of the notion of assemblage builds on framings in anthropology and geography that use it to map historically contingent convergences (Ong and Collier, 2005; Marcus and Saka, 2006; Li, 2007a; Anderson et al., 2012; Dittmer, 2014) and the ways they often coalesce in objects, spaces, and landscapes (Braun, 2005, 2006; Ogden, 2011; Ranganathan,

2015; Smith and Dressler, 2017). We use it not as a means of rejecting history, political economy, or biopower but rather to trace particular possibilities at specific moments and places. We offer readings that inquire into the material and discursive, human and non-human agencies involved in shaping connections between the often heterogeneous elements at play in the making of frontier space. Writing of conjuncture, a framework to which our notion of assemblage shares significant resemblance,[4] Tania Li writes, 'Rejecting notions of a functional equilibrium, a conjunctural approach treats practices that appear to hold constant for a period of time as a puzzle, as much in need of examination as dramatic change' (Li, 2014a: 18; see also Müller, 2015). Read as such, the examination of assemblages offers a way of understanding the temporalities and spatialities of configurations of frontier opportunity, value, and violence.[5] As Li further notes, within these assemblages 'elements are drawn together … only to disperse or realign, and the shape shifts according to the terrain and the angle of vision' (Li, 2007a: 265). Assemblage thus provides a 'frame of specific complexity around the vision of unstable, heterogeneous structure' (Marcus and Saka, 2006: 104). It directs us to understanding the social world as transitory, mosaic, and fluid and helps us understand or decipher the messy interactions between new strategies of capital accumulation and the politics of space and place in frontier zones (Massey, 1994). And perhaps most centrally, it offers a non-deterministic frame for thinking through the shifting temporalities, interests, materialities, and imaginations that cohere at particular moments to produce particular spaces as resource frontiers.

If assemblage's history is fairly short, the notion of the frontier has a longer and more ambiguous trajectory. The term has been widely, and often unreflectively, applied as a heuristic device to describe processes of transition, exclusion and inclusion both physically and figuratively. There are myriad ways to approach the subject and a lack of anything resembling conceptual consensus has made defining the concept a challenging endeavour. The concept of frontier first emerged in Europe in the fourteenth century with the French word 'frontière' indicating a façade in architecture. Only later did it come to mean the limits of state control or edge of empire (Rieber, 2001: 5812; see also Febvre, 1973). There is an intimate, but often unclear, relationship between the word and the concept of frontier (Febvre, 1973). Within the English and American tradition this is further complicated by the use of the word interchangeably to denote literal borderlines, figurative borderlands, regions just beyond the pale of settled areas, and the process of territorial expansion of state authority or civilization into remote 'wastelands' and margins (Wendl and Rösler, 1999; Brown, 2010). As Redclift argues, 'The frontier is both a boundary *and* a device for social exclusion, a zone of transition *and* new cultural imaginary' (Redclift, 2006: viii). Frontiers often refer to regions where the state is presumed to struggle to assert its authority and is thinly spread (hence the notion of 'frontier justice' as a form of violent rule that is negotiated at an eminently local level). But frontiers are also liminal spaces open for production and inventiveness (Korf and Raeymaekers, 2013). They can be zones of Schumpeterian creative destruction and transformation where imagined wastelands and backwaters presented as unoccupied and vacant are turned into sites of capital accumulation. At the same time, they may also become spaces of social experimentation, innovation and hybridity where new political subjectivities are shaped and new governance structures tested.

For our purposes, we understand frontiers as *imaginative* – zones in which the material realities of place are inextricably bound to various visions of and cultural vocabularies for what the frontier is and might be. To that end, three classic framings of the frontier are useful to keep in mind as we develop our argument. These framings are not objective realities or ideal types of frontier spaces. Rather, they are imaginations of the relationship between margins and centres that, as authors in this volume show, have grave and eminently tangible implications for these spaces and those living within them. First, the frontier is often imagined as integral to broader economic activity. The frontier has historically been framed not only as a space of entrepreneurial opportunity, but as a zone that is fundamental to the survival of capitalism itself. This argument can be traced through any number of theories of political economy, but is particularly evident in the Marxian tradition. Here, the frontier figures alongside of surplus-value as the (other) engine of capital. Frontiers are the condition of possibility for capitalist expansion (Patel and Moore, 2017). They are the sites in the midst of incorporations and enclosures through what Marx called primitive accumulation (1976) and what Harvey reframes as accumulation by dispossession (2005). Here, frontiers are sites where capitalist crises of over-accumulation are resolved in ways that forestall broader crises and systemic collapse. In other words, frontiers are imagined to be the necessary historical counterweights to industrialization. Our suggestion in calling this an imagination is not to take issue with Harvey. Indeed, in many of the sites explored in this volume, accumulation by dispossession seems an apt descriptor. Rather, it is to call attention to a particular vision of frontiers that reduces them to functions of capital. This vision of the frontier, we suggest, animates a broader understanding of frontiers as crucial spaces that hold the key to economic expansion, development, and growth.

The second is the anxious imagination of the frontier as a site of danger and lawlessness. Here, frontiers are construed as the rims of empire – spaces at the limit of the reach of state control. The frontier is the literal margin of the state, a zone ambiguously within state power, where law and order are limited, loyalties are questionable or ephemeral, and security is at once crucial and tenuous (Das and Poole, 2004; Tambiah, 2013). A classic articulation of this imagination is Lord Curzon's famous *Frontiers* lecture (Curzon, 1907). Written against the backdrop of the Great Game for empire in central Asia, Curzon, the Viceroy of India from 1899 to 1905, imagines frontiers as spaces that are little understood even as they are at the heart of colonial politics. Curzon argues that frontiers are potential instruments of rule – spaces that, if they *could* be demarcated, managed through sound colonial policy, and settled, might prove as bulwarks against chaos. Yet, his treatise simultaneously identifies them as spaces that are unquiet and hard to govern. He thus imagines the deployment of political technologies of territorial rule (Elden, 2013) as able to transform these spaces from sites of imperial anxiety to instruments of imperial peace. At the same time, his lecture points to the limits of such technologies for settling imperial space. Curzon's piece is thus a paradigmatic framing of the 'unruly frontier': a zone that is once ungovernable and in urgent need of government, a space that needs territorial incorporation but might be unincorporable.

The third is the imagination of the frontier as a wilderness space: untouched, unpeopled, and open for exploitation. This framing is resonant with Frederick Jackson Turner's 'Frontier Thesis'. Turner's well-trod and critiqued argument describes the influx of European settlers into the American 'wild west' in search of 'free land' as transforming spaces of wilderness and vastness into spaces of civilization and order. For Turner, this was part of a short-lived linear process of exploitation, conquest and pacification that steadily would engulf the wilderness and its primitive population, therein closing the frontier. In the words of Turner the frontier was thus the 'meeting point between savagery and civilization' (Turner, 1920: 3). Turner's writing describes a specifically North American process. Yet, it resonates with a much broader contemporary and historical pattern of representing frontiers in terms of discovery, emptiness and underuse – a trope that legitimates 'progressive' development and control of the margins (Li, 2014b). As Anna Tsing notes, 'frontiers create wildness so that some – and not others – may reap their rewards' (Tsing, 2005: 27). In other words, part of the acme of frontiers are their often-fictive framing as unpeopled wilderness, pregnant with possibility and open for intervention.

Though careworn, these imaginative tropes of frontier thinking remain key ways of framing marginal spaces – borders, borderlands, upland areas, remote forest zones, deserts, steppes, coastal hinterlands, 'waste' or 'idle' zones – as resource frontiers. They help imagine these spaces as simultaneously critical, open, and in need of intervention.

Resource Frontiers as Frontier Assemblages

Based on our collective observations of contemporary territorial transformation across Asia, we believe an alternative mode of imagining – and analysing – frontiers is in order. The analytic of frontier assemblages provides a way to interrogate the contingency, emergence, rupture, possibility, and visions of modernity at work in these projects. Yet, equally importantly, is to show how these spaces emerge as new laboratories of socio-economic, ecological, and spatial ordering (Tilley, 2011; Cons, 2018). We set ourselves the task of not only understanding who wins and who loses in these new configurations, but also how the conditions of success and failure emerge in often surprising and contingent ways. We enquire into the ways that resource frontiers become strategic spaces for assembling land as a resource for investment (Li, 2014b). Moreover, we ask about the broader processes at work that precede, constitute, and follow the assembling of resource frontiers as epicentres of extraction and production.

In developing this analytic, we build on the long tradition of work on resource exploitation in political ecology (see for example Peluso, 1992; Peet and Watts, 1996; Peluso and Watts, 2001; Heynan et al., 2007; Perreault, Bridge, and McCarthy, 2015) and rich work within this tradition that has begun to map new resource frontiers in Asia and beyond (Fold and Hirsch, 2009; McCarthy and Cramb, 2009; De Koninck et al., 2011; Laungaramsri, 2012; Hall, 2013; Bennike, 2017; Rasmussen and Lund, 2018). Existing work has provided insight into the complex, multi-scalar factors that shape these interventions and expansions. It also sheds light on the relationships between the complexities of place and broader geopolitical transformations that are increasingly central to the economic and political agenda of many Asian states (Sturgeon, 2005; De Koninck, 2006; Barney, 2009; Hirsch, 2009; Lund, 2011; Woods, 2011a, b; Laungaramsri, 2012; Levien, 2012; Cons and Sanyal, 2013; Lagerqvist, 2013; Eilenberg, 2014). This research has outlined a set of crucial dynamics for understanding contemporary frontier assemblages and their histories. Drawing on the case of Laos, Barney, for example, argues that state agencies utilize the discourse of 'the last frontier' as a strategy for attracting transnational investment and legitimating the conversion of its uplands into capital-intensive resource extraction zones. Barney here applies the notion of the 'patchworked frontier' to frame the relations between new global investments and previous regimes of resource governance that produce overlapping mosaics of regulation and control (Barney, 2009: 147). Such 'frontier neoliberalism' is a cyclical phenomena that waxes and wanes according to the strength of the state and the pressure of global markets (Wolford et al., 2013a). Büscher reiterates that, 'frontiers have special significance in a neoliberal political economy. Neoliberalism needs frontiers ... neoliberal capitalism thrives on frontiers' (2013: 10). However, these new expansions are instruments not just of profit, but also of making troublesome spaces legible, manageable, and secure. State-backed funding (and military security) for private investors has often created reassurance for large-scale investments and new business opportunities in these often politically contested spaces. Moreover, as contributors to this volume show, the current round of frontier making in Asia cannot narrowly be described as the outcome of neoliberal processes (see Rubinov; Zee; Choi; Woodworth; Lentz, this volume). Many of the frontier zones under examination herein emerge out of socialist and post-socialist histories and trajectories that resonate with neoliberal projects even as they complicate our understandings of their meanings (Collier, 2011).

Broadly across Asia, resource frontiers are emerging as spaces of legalized lawlessness where aleatory forms of sovereignty determine who qualifies for citizenry and who will be excluded from the nation-state project (Dunn and Cons, 2014). We see these accelerated processes of dispossession in states such as Myanmar and Indonesia where resource frontiers are often militarized spaces for control and extraction (Woods, 2011a; Eilenberg, 2014). Military involvement in resource extraction and land dispossession in Asian frontiers as a whole is not a novel phenomenon. It can be traced back to the counter-insurgencies of the Cold War era where many of the burgeoning Asian nation-states were plunged into violent conflict, instigating processes of forced resettlement, resource exploitation and firm military control (De Koninck, 2006; Peluso and Vandergeest, 2011). Many of these frontier zones have since been under various forms of military authority, often becoming zones for economic exploitation generating revenue for military budgets. As noted by Rasmussen and Lund, 'Frontier spaces are where the often violent destruction of previous orders take place, and the territorialization of new orders begins' (2018: 396). In this sense, converting marginal spaces into resource frontiers has codified state power in these unruly landscapes and strengthened control of areas that often trouble territorial sovereignty. Both in their past and present iterations, their reframing as resource frontiers heralds a new wave of interventions within them.

The Dynamics of Frontier Assemblage

Frontier assemblage draws attention to the lifecycles of these frontier spaces, the dynamics of their emergence, their unravelling, and their aftermaths. While these dynamics fluctuate in ways that defy simple causal explanations, they do share characteristics and resonances that demand comparison. In the remainder of this introduction, we frame a set of dynamics that contributors show are key to, but often obscured in, processes of making new resources frontiers.

To do this, we suggest decentering, though not eliminating, a focus on financialization in the making of frontiers. The financialization of frontier space might be thought of as, following Randy Martin, a process whereby socio-spatial relationships and affiliations are 'reconfigured to extract wealth as an ends by means of risk management' (Martin, 2007: 7; see also Roy, 2012). Understood this way, resource frontiers appear as key sites in a broader terrain of what Ananya Roy calls *riskscapes* – formations of territory organized around broader technologies of managing accumulation and risk (Roy, 2012). In this broader territory, resource frontiers are risky sites of risk management – spaces in which broader crisis is mitigated through extraction carried out in often short-term, capital intensive, and speculative ways (see contributions to this volume by Anderson and Paprocki). Yet to understand the formation of these riskscapes, we suggest, requires understanding financialization, social transformation, politics, and ecological change as mutually constituted with and in resource frontiers. To that end, we trace three linked dynamics that are intimately connected to financialization but often take surprising relationships to it.

Frontierization

Frontiers are mutable, temporal, and mobile entities (Cronon, 1996). They emerge at particular conjunctures and disappear at others. They have lifecycles, deaths, and occasionally, peculiar rebirths (Geiger, 2008; Korf and Raeymaekers, 2013). Moreover, they do not have fixed boundaries. From any one place, a 'frontier' can bleed out, expand, and contract. The frontier can move on and return. At the same time, the frontier might be resurrected in new forms that build on the ruins of others (see Lentz, Middleton, and Rubinov, this volume). Such dynamics often serve the needs of capital. But their emergence hinges on a broader range of techniques than land acquisition and investment alone. Indeed, of crucial import here are the temporalities of frontiers, the ways that they form and deform at particular conjunctures and disjunctures. As contributors to this volume show, understanding the framings and potentialities of frontier space requires thinking equally about frontier time.

Anna Tsing argues that a frontier is 'an edge of space and time: a zone of not yet – not yet mapped, not yet regulated. It is a zone of unmapping: even in its planning, a frontier is imagined as unplanned.... Their wildness is made of visions and vines and violence: it is both material and imaginative' (Tsing, 2003: 5100; see also Bridge, 2001). Tsing's evocative description highlights a need to think beyond a narrow framing of frontiers as places where resources are discovered and subsequently exploited. Rather, it underscores what we see as a need to understand the processes of making the frontier as eminently entangled: anchored in the imaginative, the material, the known and the unknown.

What might this mean for investigations of 'frontier' assemblages? As various authors have shown, a critical element in making the frontier is the framing of these spaces as timeless, unpeopled lands – open to extraction and exploitation of various sorts. This suggests that frontierization must be understood as a process of radically simplifying the meanings of a space to, primarily, the things valued within it. This simplification implies that the relationship between resources and spaces is anything but incidental. Rather, in resource frontiers, resources and frontiers are co-constituted. Yet, this relationship and the techniques on which its production hinge, are highly unstable. Making the frontier, as Tsing notes, is an unmapping of place – a reduction and elimination of the dynamics that constitute a specific locale through a process of rendering extractable: transforming space and place into land and property ripe for exploitation. Accomplishing such feats requires a range of representational and inscriptive technologies – maps, satellite images, fences, property titles, etc. (Li, 2014b). Yet, the relationship between reality and its representation is never an innocent one. As Timothy Mitchell notes, the mapmaker, or the inscriptive technology more broadly, 'cannot keep reality out of (its) representation' (Mitchell, 2002: 116). Landscapes and the people, flora, and fauna within them are active participants in the making of the frontier, even as they are rendered invisible or as technical problems to be managed through discursive and material force. The notion of the frontier then might best be thought of as an enframing strategy whereby space is rendered as an extractive territory temporarily open for frontier management (Mitchell, 1991; see also Heidegger, 1977). Yet resource frontiers are spaces that regularly fail to conform to such renderings. These various failures constitute the terrains of breakdown, contestation, and conflict that often characterize resource frontiers in contemporary Asia. The things erased in the making of the 'frontier' return: sometimes as hauntings, sometimes as unanticipated dynamics which foil neatly laid plans, sometimes as violent resistance.

An analysis attentive to the ways that landscapes themselves are entangled in the making of the frontier, opens a set of critical questions for understanding frontier assemblages (Ogden, 2011). It points towards a rethinking of the materiality of terrain and a necessity to see violence not only as the byproduct of capital (Sivaramakrishnan, 1999; Elden, 2010). Yet, it also focuses our attention on the outcomes and effects of the frontier itself, and the ongoing questions of life in the midst of ruination (Stoler, 2013). Resource frontiers, precisely through their 'productive' capacities, produce ruins (Choi, this volume). They denote extractions from landscapes in ways that often irrevocably transform them. Frontiers leave land in their wake in which the relationships between landscape and the people, animals, and plants found within them might be permanently transformed (Hall et al., 2011). Moreover, the kinds of transformations wrought by extractive projects have unfolding and ongoing effects – whether through pollutants and toxicity, the integration of margins into the fold of governmental power through new infrastructure, or the permanent alteration of land and property relations (see Choi; Woodworth; and Zee, this volume). As Gaston Gordillo notes, attending to the rubble that remains in the wake of such projects 'helps us understand the ruptured multiplicity that is constitutive of all geographies as they are produced, destroyed and remade' (Gordillo, 2014: 2). To understand frontierization and its production of mobile, mutable frontiers thus demands attention to what happens when frontiers move on and, as Kasia Paprocki argues here, to the ways ruination itself is imagined as an object and opportunity in planning schemes.

Remoteness/proximity

The concept of a 'frontier', not dissimilar to that of a 'colony', articulates a relationship of spatial control. The frontier is a zone at once 'in relation to' and 'at a distance from'. It exists in dialectic tension with metropoles and centres of various sorts (Stoler and Cooper, 1997). The production of the frontier as an anomalous zone is a relation of distance that licences certain forms of experimentation within it (Sivaramakrishnan, 1999). This presumed distance is precisely what characterizes and legitimates intervention and redistribution within it (Swanson, this volume). Yet, frontiers often prove to be much more proximate than they are imagined – whether as spaces which become intimately linked to metropoles through migration, through atmospheric displacements such as those discussed by Jerry Zee in this volume, or through infrastructures such as pipes and roads. Indeed, as urbanization driven growth models increasingly characterize expansion around Asia, the distinction between frontiers and centres further blurs (see McDui-Ra; Choi, this volume). The imaginations of distance often give way to realities that tie the frontier tightly to the space from which it is framed as remote (see Günel, this volume). Thinking through this dynamic tension, we suggest, is productive for understanding the ways that frontier assemblages emerge.

Particularly instructive here is a classic essay by anthropologist Edwin Ardener which posits a phenomenology of remote areas. For Ardener, the relations of remoteness must be understood as a social relation intimately bound up in power. As he writes, 'The actual geography (of a remote area) is not the overriding feature – it is obviously necessary that "remoteness" has a position in topographical space, but it is defined within a *topological* space whose features are expressed in a cultural vocabulary' (Ardener, 2012: 532). For Ardener, the geography and geometry of remoteness are fundamentally relational in character, constituted by the social and cultural configurations and imaginations of territory. This topology articulates a relationship whereby spaces might be simultaneously imagined as peripheral and central to national, and other kinds, of territory (Cons, 2016, 2018). Remoteness, as such, is a dialectical formation that produces possibilities, opportunities, and relationships of power. As Harms and Hussain write in a recent commentary on Ardener, 'remoteness is not so much a place as a way of being' (2014: 362; see also Mathur, 2016).

We suggest that this socially constituted relation of remoteness is a fundamental characteristic of resource frontiers, one that contributors to this volume are at pains to elaborate (see Günel; Middleton; McDuie-Ra; Rubinov; Woodworth, this volume). While these dynamics shift from place to place, remoteness describes a particular landscape of opportunity, illicit intervention, and regulatory power. Ardener enigmatically writes that 'remote areas turn out to be like gangster hideouts – full of activity and half-recognized faces' (Ardener, 2012: 524). Another way of saying this is that the notion of a remote area articulates an imagination of space and the people residing within it as on the margins of the pale – as living at the limits of state power and making a living out of ambiguous relations to it. This ambiguous liminality of frontier space demands an attention that traces not only the social production of remoteness, but also the tensions, anxieties, and recursive interventions that remoteness heralds.

Overlapping rule

Our final point of departure for frontier assemblages is to build on longstanding critiques of the notion of state, sovereignty, and territory to foreground frontier spaces as especially brittle and contingent parts of the mask of state territorial control (Abrams, 1988). Frontiers emerge as spaces where the veneers and contradictions of rule are often particularly apparent. This is not to subscribe to the imagination of frontier spaces as a fault line between chaos and control and between rule and lawlessness. Rather, it is to recognize that they are spaces in which multiple interests, bids for sovereign control, and attempts to monopolize opportunity and access often accumulate (see Anderson; McDuie-Ra; Paprocki, this volume). Such accumulations are further complicated through the dialectics of official attention and neglect that often characterize margins throughout Asia (Cons, 2016).

Frontier assemblages thus often emerge out of longstanding contestations and combinations of sovereign and territorial power (Lentz, Swanson, this volume). They are spaces within which a multiplicity of interests and forms of rule accumulate and overlap to produce spaces that are *anything but* unruled, even if they may fail to cohere to state, corporate, transnational agencies or other sovereigns' logics of legibility, order, and control. Rather, the patterns of governance and rule that emerge within frontier spaces often have an aleatory character to them, a sense that rule is unknowable, unpredictable, and often accomplished by chance (Dunn and Cons, 2014). In other words, frontier assemblages are rarely projects of successfully shoring up uniform and coherent governance. More often, they are characterized by multiple interventions seeking to order land, people, and nature that unfold in ways that are often indeterminate, both for those who live within these spaces and for those who seek to govern them. Sometimes patterns of overlapping rule in frontiers produce coherent regimes of extraction, exploitation, and opportunity. Sometimes they contest with and undermine one another. Sometimes they produce surprising and unintended results which foreclose on old opportunities while evolving new ones.

These contingent configurations should not be understood as failures to develop regimes of sovereign control, but rather as, themselves, the frameworks of rule within frontier space. Christian Lund argues that patterns of control over land should not be understood as reflecting pre-existing conditions of sovereignty and power. Instead, they produce it. That is to say, the overlapping claims to control that often characterize frontier space should be understood as assembling particular, albeit fragmented, forms of sovereignty and rule (Lund, 2011). Frontier spaces thus trouble understandings of sovereignty as a project of interiorization (Deleuze and Guattari, 1987; Agamben, 1998). Here, they emerge as ambiguous zones that are neither clearly and coherently within nor outside the logics of territorial rule. This ambiguity is often complicit in provoking territorial and cartographic anxieties over the nature of frontier land, of the resources in it, and of its residents (Krishna, 1996; Billé, 2016; Cons, 2016). A critical task of the analyst of frontier assemblages is to understand how these fragments do, or do not, cohere (McDuie-Ra, 2016).

Conclusion

The chapters that follow trace the formation of frontier assemblages across Asia. We seek to chart a set of resonances and processes that are at once hyper and trans local. Our approach to understanding these assemblages is not to chart out an exhaustive network of factors that make frontiers, but rather to illuminate a series of often overlooked processes and imaginations at work in the assembly of new resource frontiers in Asia. Authors in this volume contribute studies from diverse regions across the continent. While far from exhaustive, we here offer explorations of frontiers in South Asia (McDuie-Ra, Middleton, Paprocki), South-East Asia (Anderson and Lentz), West Asia (Günel), Central Asia (Rubinov), and East Asia (Choi, Swanson, Woodworth, and Zee). In doing so, we offer a theoretical and methodological experiment in bringing radically different resource frontiers into the same analytic frame. In doing so, we hope to demonstrate the productive value in understanding resource frontier assemblage as a set of loosely linked processes that are happening within and across space.

This book is conceived as a dialogue between scholars of Asian frontiers and margins. It is divided into four thematic sections – experimentations, cultivations, expansions, and (re)assemblies – tracing different dynamics unfolding in and through frontier assemblages in Asia today. These thematic sections index various different processes that are at once historical and are increasingly mobilized in the production of frontiers today. Each section is introduced with a brief framing essay that situates the theoretical and empirical projects of each chapter against the broader literature on and historical trajectory of frontiers, resources, and resource frontiers in Asia.

The first section, 'Frontier Experimentations', traces a set of novel reworkings of frontier space to facilitate new kinds of management and control, particularly in the face of climate and environmental change. It charts the production of frontiers through anticipatory imaginations of climate ruination in Bangladesh (Paprocki), the reinvention of the subsurface as a frontier of carbon storage in contemporary climate change debates (Günel), and the logics of managing dust in urban areas such as

Bejing through sedentarizing and retraining mobile populations in central Asia (Zee). The second section, 'Frontier Cultivations', maps the ways that cultivation and growth are bound up in resource frontier processes both old and new. It examines logics of cultivation and care in resource frontiers through new Green Economy policies for managing climate change in Indonesia (Anderson), social practices of grafting fruit trees in post-frontier Kazakhstan (Rubinov), and the logics of salmon cultivation and redistribution in Hokkaido, Japan (Swanson). The third section, 'Frontier Expansions', traces the articulations between urban growth and frontier space, examining the ways that urbanization is increasingly, if paradoxically, a central trope in Asian resource frontiers. It traces this expansion and incorporation through the development of new Chinese eco-cities on land reclaimed from the ocean (Choi), monuments and ghost-towns in urban peripheries in China (Woodworth), and the emergence of private healthcare markets in Imphal, India (McDuie-Ra). The final section, 'Frontier (Re)Assemblies', questions the after-lives and reinventions of frontier space. It does this through an examination of the long and unfolding history of cinchona plantations in India's Darjeeling region (Middleton) and the subsumed histories of a contemporary hydro-electric resource frontier in Vietnam (Lentz).

This volume offers a set of what Prasenjit Duara has called convergent comparisons that illuminate emerging dynamics reconfiguring relations to land and accumulation within Asia and beyond (Duara, 2016). The book interrogates this unprecedented transformation and the complex array of actors, forces, and ecologies that constitute it. That said, our notion of assemblages does not suggest a random or incommensurate coming together of these relationships. As Li writes, 'a conjuncture [or an assemblage] isn't radically contingent: all of the elements that constitute it have histories and there are spatial configurations that make certain pathways easier or more difficult' (Li, 2014a: 150). We see these relationships and their histories as the starting point of critique – a set of processes in need of demonstration and elaboration if one is to understand the ways that contemporary and historical Asian resource frontiers continue to dramatically shape politics, socialities, ecologies, and economies.

Notes

1 See for example (De Koninck et al., 2011; Pye and Bhattacharya, 2012; Fox and Castella, 2013; Verbrugge, 2015; Peluso, 2018).
2 In referencing Tsing's (1993) notion of an 'out-of-the-way place', we invoke her understanding of marginality as not simply a geography but a set of social and political relations with long and often violent histories. Margins and marginality, for her, are cultural and political constructions. These

constructions are 'about the process in which people [and we would add places] are marginalized as their perspectives are cast to the side or excluded. It is also about the ways in which people actively engage their marginality by protesting, reinterpreting, and embellishing their exclusion' (1993: 5). The notion of margins and 'out-of-the-way places' articulates well with Ardener's (2012) discussion of remoteness, which we build on later in this chapter.

3   Such an approach has proven particularly productive for scholars engaged in Actor-Network Theory and the post-human turn because the move towards assemblage opens the possibility for a rethinking of agency – one which is not intelligible within enlightenment epistemologies organized around the binary of nature (as passive non-actant or stage) and culture (humans as the sole and heroic drivers of historical change) (DeLanda, 2006; Latour, 2007; Bennet, 2010; Müller and Schurr, 2016).

4   Conjuncture and assemblage draw from different theoretical traditions (Gramscian and Deleuzian respectively), and thus harbour markedly different epistemic positions. Yet, following work in the anthropology of the state, we argue that keeping these perspectives in productive tension with one another offers broader perspectives on the makings, ambiguities, and temporalities of the frontier (Hansen and Stepputat, 2001).

5   The notion of assemblage that we are working with here bears similarities to Althusser's notion of a 'complex whole', where multiple temporalities come together to produce a time of times that 'cannot be *read* in the continuity of the time of life or clocks, but has to be *constructed* out of the peculiar structures of production' (Althusser et al., 2015: 248). Our thanks to Vinay Gidwani for pointing this out.

Part I
Frontier Experimentations

Framing Essay
Assemblages and Assumptions

Christian Lund

Is there more God in an elephant than in an oyster?
— Samuel Beckett, *Echo's Bones*, 1933

The human understanding is of its own nature prone to suppose the existence of more order and regularity in the World than it finds.
— Francis Bacon, *Novum Organum*, 1620

Assemblages have gained currency as an object of study as well as a method of inquiry. The Introduction to this book by Cons and Eilenberg shows how objects are conceptualized as complex assemblages by a variety of disciplines and how it can be used to analyse specific historical constellations of dynamics. There is something intuitively appealing about looking at the world as assemblages. The empirical world is messy, and one is well-advised to be open to the multitude of actors, resources, objects, ideologies and discourses without assigning over-determining qualities to any of them prior to analysis. The world can, indeed, be seen as an infinite number of oysters and elephants making up a transitory mosaic of facts. Openness is a sound scholarly reflex, but we should not claim to be more open-minded than we can be. It can be tempting to simply 'let the naked facts speak for themselves', yet, no naked facts ever spoke to anyone except through concepts and assumptions, however vaguely defined or unacknowledged. In the social sciences, the potential variables are simply too many for us to not select

Frontier Assemblages: The Emergent Politics of Resource Frontiers in Asia, First Edition. Edited by Jason Cons and Michael Eilenberg.

and specify a few. We select all the time, and we do it through concepts and assumptions; through epistemology. The transition from observation to analysis, is not like passing through the green gate in an airport with 'Nothing to Declare'. We always travel with our epistemology, our concepts, and our assumptions. To not declare them does not mean that you do not have any. But by not questioning the concepts and categories with which we read the 'naked facts', and by not making our assumptions clear and laying them out, it is easy to make a set of observations look speciously unequivocal and pervasive (Lund, 2014: 226). We always have intellectual preferences about what pickings of the magpie's nest of reality we find important and can foreground, and what knick-knack we should disregard for the moment. It is imperative to acknowledge that the inspection of an assemblage is inevitably guided by assumptions. And they should be made explicit. Declare the goods.

This line of thinking has roots in Immanuel Kant's work (Kant, 1953; Wolf, 1999). He argues that rather than insisting that our cognition reflects 'real' objects, we may try to see 'real' objects as conforming to our cognition. That is, we 'see' things through a lens of a priori concepts formed through experience; these newly observed things then become part of our new experience, and we can have another, new, look. This suggests a never-ending, iterative approximation between a priori concepts, cognition of 'the world', and the formation of renewed a prioris. Hence, the transition from what we *see* to what we *understand* requires open conceptual determination; maybe especially with assemblages. In the Introduction to this book, for example, Cons and Eilenberg focus on materiality, not for materiality itself but for its connection to, say, institutions. They focus on institutions not for institutions themselves but for their connection to, say, ideas. They focus on ideas, not for the ideas themselves ..., and so on.

In her work on anticipatory ruination in Bangladesh, Kasia Paprocki identifies a series of dynamics with clear assumptions about how materiality and its institutions link with ideology, even when, in the case of the World Bank, it is packaged as 'robust options' with a sheen of scientific objectivity. Paprocki shows how governments and other influential policy makers perform so-called epistemic interventions. By framing large climate-based threats as inevitable, active, 'pre-emptive', destruction of people's living conditions is simply to give history a helpful push. Hannah Arendt shows in her writings that reading ideology as 'natural law', or 'natural history', of a teleological movement opens for terror, where the 'political' yields to 'emergency' and 'the inevitable' (Arendt, 1979). By using her concepts, Paprocki abstracts and edits her data, and she makes particular inherent qualities of it prominent.

Paprocki's identification of pre-emptive ruination suggests that climate change has the potential to sire nasty ideologies. Not only climate-change deniers will use their denial as convenient excuse; those who acknowledge climate change can equally use this to justify and expedite atrocities when opening new frontiers.

In her work on sub-surface frontiers, Gökçe Günel explores the efforts to exploit the underground; not to extract resources but to put waste back in, more specifically carbon dioxide. This is a complex technical operation with equally complex questions of law, ownership, risk, and interests. The analysis focuses on the discursive and accountancy work that goes into reconstituting the underground to become a space for burying the excrements of the carbon-based economies. Günel makes the assumption that not only are frontiers assembled, but are multi-local. The site of accountancy work is not in situ of the depositories, but more importantly, the storage will increase the relative value of other resources very far away (where it would otherwise have been dispersed, dumped, or deposited). This opens for an important complementary perspective to existing works on the carbon frontier (Mitchell, 2011; Watts, 2014), and the striking feature is the confluence of interests. This new frontier of carbon dioxide storage, has many resemblances with the extraction of the carbon in the first place. One reason is the technical and legal nature of the operations, another that storage of carbon dioxide makes the emission problems 'go away', and can therefore be used to justify continued carbon dependency.

In his work on governance and the Chinese dust-bowl, Jerry Zee deals with an extraordinary frontier. Contrary to conventional frontiers that mark the expansion of capitalism and the destruction of rights and institutions pushing ever further out, this frontier rolls the other way. Zee makes the assumption that even when certain frontier dynamics are halted, it makes sense to see this slow-down as the result of the acceleration of other, competing frontier dynamics valorizing new resources, new activities, and in the end, new actors. Dust storms and desertification in Inner Mongolia appear to move closer to Beijing and threaten the Chinese heartland. So, to combat this threat, the government authorities engaged in complex measures to control the environment as part of their larger project to create a 'socialist ecological civilization'. This policy reworks the frontier regions of China. The 'environment' is, in Zee's text, a feat of symbiotic engineering where government works through logics and incentives to have herders and others replant, stabilize and re-establish the landscape. The different subject positions that government policy animates are far from stable, however. The efforts to roll back the physical frontier of sandstorms and stabilize the environment may, thus, also open a new frontier of destabilization of the social environment, labour, and belonging. A destabilization that could surpass the damage of storms.

The three contributions work on new frontiers and show how to stretch the concept. 'Social scientists do not discover new events that nobody knew about before. What is discovered is connections and relations, not directly observable, by which we can understand and explain already known occurrences in a novel way' (Danermark et al., 2002: 91). Paprocki, Günel, and Zee encourage us to see how new valuable resources emerge in changing contexts. Climate calamities, such as flooding, sand storms, and global warming, produce new emergencies that can override the political. And they open for the use of resources as varied as the land of the evicted/evacuated, the underground for geological carbon sequestration, and the labour of herders-turned-into-foresters and a desert which may, one day, be fertile. These situations do not conjure up the usual images of frontiers as the wild west and un-conquered expanses. Instead, they demonstrate how frontiers represent, most basically, the discovery or invention of new *resources*; not new *places*. They may be in our back-yard, hell, they may *be* our back-yard. As when the municipality of Copenhagen 'discovered' an 'unused' marshland behind a now redundant industrial area and decided that hover-flies, green toads, and Copenhageners' Sunday outings, were standing in the way of profitable condominiums filled to the brim with new tax-payers who could fund a metro line. Consequently, a 'frontier is not space itself. It is something that happens in and to space. Frontiers *take* place' (Rasmussen and Lund, 2018: 388). This 'something that happens' which turns the potential into an actual resource, is complex. It is a complex of material resources that are restructured by actors who wield power and knowledge. They hold the capacity to frame the potential in narratives of prosperity, and security, and then undergird their interests with ideology and force, law, and exception (Mattei and Nader, 2008). Such complexes have structure, contours, and topography: The world is not flat. Messy, yes; flat, no. The social and political landscapes that emerge from these works are far from the hodgepodge of 'everything is connected and matters equally'. Instead, the three authors' concepts illuminate the contours of the frontiers and show the unfolding of events as contingent necessities (Jessop, 1990) rather than pure accident or structural pre-destination. Through their epistemological experimentation, the contributors open up new vistas for the understanding of the rocky resource frontiers.

1

All that Is Solid Melts into the Bay

Anticipatory Ruination on Bangladesh's Climate Frontier

Kasia Paprocki

What are the ruins of the bourgeoisie? – Walter Benjamin (1999)

Bangladesh's southern coastal region is home to the world's largest delta, that of the Ganga, Meghna, and Brahmaputra rivers, that drains into the Bay of Bengal. This region, and particularly the southwestern district of Khulna, is often referred to as the most vulnerable to climate change in the world.[1] Key actors in policy, academic, and practitioner communities have raised the alarm over the threat of rising waters to communities that inhabit the low-lying coastal islands, which trail off into the Bay across the deltaic plain. In response, many within the government and donor community have identified shrimp aquaculture as a principal adaptation strategy throughout the coastal landscape (Paprocki and Huq, 2017). The production of shrimp for export transforms the threat of rising sea levels into a market opportunity. Shrimp ponds now dot Khulna's coastal landscape; framed by notions of resilience and adaptation, donors and development agencies are re-shaping the ways in which Khulna's landscape is understood, managed, and transformed.[2]

The shaping of Khulna as a frontier of climate change adaptation is an active and ongoing process, involving both epistemic and material dynamics. The mobilization of shrimp aquaculture within this climate frontier is not incidental to it; rather, shrimp production plays a critical role in shaping the formation of the frontier. In particular, shrimp aquaculture

Frontier Assemblages: The Emergent Politics of Resource Frontiers in Asia,
First Edition. Edited by Jason Cons and Michael Eilenberg.
© 2019 John Wiley & Sons Ltd. Published 2019 by John Wiley & Sons Ltd.

in this landscape is integral to the dynamics of what I call *anticipatory ruination*, a discursive and material process of social and ecological destruction in anticipation of real or perceived threats. Anticipatory ruination is a process that happens at and produces the climate frontier in Bangladesh, and is thus a fundamental dynamic of the frontier assemblage of this region. I use the concept to explore the dynamics taking place in Khulna in particular, and also the ways in which frontier assemblages are dialectically constituted (epistemically and materially) more generally.

My understanding of ruination draws on the work of Ann Stoler and contributors to her volume *Imperial Debris* (2013), who explore ruination as an act, condition, and cause, each with its own temporality. Emphasizing the active and continual nature of ruination denaturalizes the dynamics of destruction: who is the agent of ruination? How, why, and when does it take place? Thus, examining ruination as an ongoing process is to investigate claims of drivers of dispossession as remote, inevitable, or complete. Stoler and colleagues also direct attention to the historical dimensions of ruins, not as dead material artefacts of destruction, but as ongoing political projects that continue to have real effects. As such, I understand anticipatory ruination in Khulna as an ongoing act with both histories and futures, linking climate change to the longer ecological history of this region: to shrimp production, land use changes, and visions of adaptation and inevitable displacement. To understand ruination in this context is to see how the possibility of climate crisis is actively produced in particular places through modes of power that operate at multiple scales.

I further draw on a growing literature that examines new forms of governing through analysis of the politics of *anticipation* (Adams et al., 2009; Anderson, 2010; Cross, 2014; Zeiderman, 2016). This literature understands *anticipation* as a key technology of policing in a new world of increasingly securitized governance (Stalcup, 2015), and identifies how anticipation precipitates certain forms of violence. As Cons and Eilenberg highlight in the introduction to this volume, anticipation is also a key dynamic in the formation of frontier assemblages. Beyond these specific literatures, anticipatory ruination also pushes forward a robust genealogy of work in political ecology that examines how adaptation discourses can obscure causality of vulnerability, preventing structural responses to social and environmental threats (Watts, 1983, 2015b; Ribot, 2014; Wisner et al., 2004).

The climate frontier is linked to Khulna's pre-existing political economy of development, of which shrimp aquaculture was already a key feature. The ruins of shrimp aquaculture introduced through previous development interventions are mobilized in the contemporary and future ruination of the region. Beginning in the 1980s, structural adjustment programs began supporting commercial shrimp aquaculture in Khulna (Adnan, 2013), with significant funding from the World Bank and the United States Agency for International Development (USAID) (see Figure 1.1). At that point, shrimp was offered as an opportunity to diversify and promote the country's export-led growth. Today, frozen shrimp is Bangladesh's second largest export after garments.

Yet, shrimp cultivation has also been implicated in ecological degradation and the dispossession of local rice farmers, and has produced a landscape that visually invokes quite literal ruination (see Figure 1.2). Owing to the saline water that is piped in to fill shrimp ponds, soil salinity levels are often so high as to not only preclude rice farming, but also to kill trees and other remaining plant life. One donor in Dhaka I spoke with in 2015 compared the landscapes that are produced by this degradation to T.S. Eliot's wastelands. The salt hanging in the air can leave a faintly dry and sticky residue on one's skin. The feeling of ruination in this space, irrespective of an analysis of its drivers and histories, is inescapable. The continued centrality of shrimp, through the shifting development landscape from the structural adjustment of the 1980s to

Figure 1.1 A billboard in Khulna advertising inputs for high-yield commercial shrimp aquaculture operations. Major development agencies such as USAID support such programs to expand and intensify the cultivation of shrimp for export throughout Khulna's coastal zone.

Figure 1.2   Shrimp ponds in Khulna.

the contemporary focus on climate change adaptation, is salient for understanding a longer history of shaping Khulna as a frontier zone.

The imagination of climate crisis is also critical to the work of anticipatory ruination. The self-conscious anticipation of climate crisis itself shapes a frontier that transforms Khulna's landscape in its image. The climate frontier involves an imagination not only of the opportunities embedded in the expansion of shrimp aquaculture, but also the erasure of other possible agrarian futures. Shrimp production becomes an opportunity as other modes of production are imagined as unviable. Anticipatory ruination works not only through the claims to possible futures through shrimp production, but also through the destruction of imaginations of alternative futures, such as the persistence of agriculture and the communities in Khulna that depend on it. The sense of inevitable crisis thus dialectically anticipates and produces ruination.

While my examination of the climate frontier is firmly grounded in Khulna's particular historical and political geography, its reach extends much more broadly. The frontier dynamics in Khulna shed light on the logic of anticipatory ruination that operates outside of any bounded space. To better understand these broader dynamics that I explore in Khulna, I extend my ethnographic scope to international discourses on new frontiers of development decision making in the time of climate change. I do this in order to understand anticipatory ruination at multiple scales. As Cons and Eilenberg explain in their introduction to this volume, examining frontier assemblages requires linking such processes not to suggest explicit causal understandings, but instead to highlight how flows and frictions across multiple scales facilitate transformation and accumulation. Collectively, this assemblage of development imaginaries and concrete transformations in control over land and resources illuminates how the logic of anticipatory ruination works, as well as how that logic is manifested in a particular place. By shifting between temporalities and geographic scales, I thus explore how the anticipation of the future actively shapes the politics of the present.

Robust Decision Making

The making of the climate frontier is carried out at a variety of sites, spatial scales, and institutional contexts. I begin with an examination of how discourses around climate change, and development planning in the context of climate uncertainty shape the climate frontier (see also Anderson; Zee; Günel and Choi, this volume). Researchers, donors and development planners and practitioners come together regularly at international conferences to discuss the available knowledge about climate change and how communities around the world can adapt to present and future changes. Such conversations facilitate and are embedded in anticipatory ruination in Bangladesh as elsewhere.

In the fall of 2014, I observed these dynamics clearly at a conference in Rotterdam, 'Deltas in Times of Climate Change', hosted chiefly by a group of Dutch government, private, and research agencies. In one workshop, a Senior Economist with the World Bank's Climate Change Group, used a board game called 'Decisions for the Decade' commissioned by the World Bank and the RAND Corporation to demonstrate a decision-making tool they developed known as 'Robust Decision Making'. The World Bank and RAND encourage policy makers to use this tool in contexts of 'deep uncertainty', such as climate change. The board game is intended to demonstrate to participants how decision makers identify 'robust options' for investments that are less precarious in the context of uncertainty. The goal is not only to shape the way that decisions are made, but to structure understandings of the future that will shape different kinds of decisions.

Participants in the workshop, a variety of NGO development practitioners, researchers, and government civil servants primarily from Europe, Africa, and South and Southeast Asia, gathered around tables in a large conference room at Rotterdam's World Trade Center. We were given game boards and handfuls of beans and red pebbles as the economist explained the rules. We were told that our role in the game, representing provincial governors and national policy makers, was 'to create a prosperous province and nation', which is sought in the game through decisions between investments in development or flood and drought protections. As we began, the economist encouraged eager game players, irreverently quipping, 'let's reward the winners, but also shame the losers!' Faulty decisions produce natural disasters, or 'crises', determined by a random roll of a die, dubbed the 'probability density function'. In each round, ersatz provincial governors for whom the 'probability density function' produced a 'crisis' were made to stand up from their seats and announce animatedly to the room: 'Oh No!' In an online video about the use of Decisions for the Decade and associated games developed by the same group, one player describes this moment in game play, 'all of a sudden a flood hit. And I died. So ... Well, I was washed away to a local slum town and have been subsisting off of left-over banana peels and whatnot.'[3] As this glib takeaway highlights, the point of the game is to allow players to imagine the kind of profound ruination from which there is no return. In the context of the game, this anticipation is pervasive – unless the player chooses the 'robust option', the threat of ruination is always present.

Moreover, the anticipation of this ruination is understood to exist without geographic boundaries. Notably, the 'provinces' we represented bore no signs of any particular place. Indeed, we were intended to imagine that they might stand in for any given province anywhere in the world. Likewise, the game sought to help us assimilate equally boundless decision-making tools, intended to be applied universally, without specification.

In the second round, a twist was introduced: instead of making decisions between standard investments in development and protection, players were offered a 'robust option', by which their investments would be protected from any climate-based variability. The economist explains, 'Say you can invest in something, for example, industry, that would guarantee [returns] without any risk of flood or drought'. So there it is, the robust option: industry. It's not vulnerable to droughts and flooding, unlike antiquated investments in things like agriculture and coastal embankments in rural areas. The economist lamented that robust options are expensive, and 'not easy', presumably a euphemism for their unpopularity with the fictitious constituencies of the game players. He exhorted sympathy (is this tongue-in-cheek?) for the 'poor World Bank that needs to hand out all the money for all these development projects to make it all robust'. Game players were inclined to agree with him, then, when he chided those investing heavily in protection that 'you're pointing your countries on a less prosperous path!' By the last round of the game, when the standard die was replaced with a floppy piece of paper taped together, labelled 'the cone of uncertainty' (apparently intended to represent the enhanced and profound uncertainty of the future under climate change), players already understood the moral. That is, ruination in the future will be inevitable; it will be necessary to make decisions that anticipate this ruination; these decisions may involve a normative shift in values, and will radically change landscapes (see Zee; Choi, this volume).

Decisions for the Decade thus facilitates the kind of 'magical vision' that characterizes frontier projects everywhere (Tsing, 2000: 133). It conjures the kind of empty, exposed space required for making decisions that appear neutral and rational. Robust options are appropriate to this new frontier, for landscapes that do not yet exist, but will. The economist explained that he and his World Bank colleagues play this very game with policy makers all over the world in order to instil these same feelings of deep uncertainty and encourage reflection on Robust Decision Making. Such tools facilitate a determination of appropriately robust decisions, involving quantitative, long-term policy analysis, and the use of modelling software to generate possible scenarios of this uncertain future.

What the board game and the discussion of the Robust Decision Making model mask is the profoundly normative nature of these decisions and the discussions informing them. They present decisions as technical where they fundamentally involve the devaluation of certain futures and livelihood strategies. Specifically, the devaluation of rural futures in favour of urban ones is a common feature of these imaginations (see McDuie-Ra; Woodworth, this volume). This devaluation is accomplished through what the economist described in a lecture he gave at this same conference as 'framing', a critical priority on which he encouraged all gathered participants to focus more. He explained that a more positive framing of the decisions about what needs to happen in anticipation of climate change is important, with the caveat that 'just this difference in framing doesn't change anything about the content', though it does facilitate cooperation. For example, while standing in front of a slide depicting an image of Mumbai's infamous Dharavi slum, he noted that 'for a lot of people in rural areas, these places are opportunities for better jobs and higher salaries, better schools for their kids, access to health care'. Framing a migration from a rural community to Dharavi, then, as an 'opportunity' for a more 'robust' livelihood is critical to this process of imagining and creating different futures. Thus, it is perhaps no surprise that reports on Robust Decision Making often cite out-migration from rural to urban areas as a key example of potential 'robust options' (Hallegatte et al., 2012; Lempert et al., 2013). As Hallegatte's comment about the rural migrants to Dharavi indicates, whether agrarian futures are 'robust' or not, they are not understood to be normatively positive outcomes of development investment.

Therefore, decisions about what *should* be done as opposed to what *could* be done are quite different. The difference reflects the normative assessments and framings of what an ideal future might look like. Robust Decision Making, as a process of anticipatory ruination, creates opportunities for certain futures, while foreclosing others. In Bangladesh, the same logic can be observed in the planning discourses concerning the future of the coastal region. I turn now to an examination of these particular dynamics, and their impacts in Khulna.

Imagined Futures

Imagining the future of coastal Bangladesh is fundamental to the process of anticipatory ruination, and is repeatedly invoked in discussions of possible interventions in the coastal zone. A frontier itself is, in Tsing's words, an 'imaginative project' (Tsing, 2003: 5102). Denaturalizing understandings of frontiers as discrete spaces facilitates an understanding of the work that the project of frontier-making does, and how it shapes landscapes through the work of imagination. Framing the region's future crisis involves claims to the need for both (i) general acquiescence and (ii) specific authority by development practitioners. Each claim is explored in turn below. These narratives claim that climate change will require certain changes, and that these claims naturalize the process of anticipatory ruination. Experts explain that while the future is uncertain, they assert the need for candid discussions about the possibility that ruination is inevitable, as well as the need to proceed in the absence of clear scientific data assessing this inevitability. Finding solutions will be difficult, as one expert explained to me, because 'all solutions are bad solutions for some people'. Thus, as another told me, echoing a common refrain among development practitioners working in this field, the adaptations that will be deemed necessary 'will make us feel uncomfortable' but 'we need to move forward despite feeling uncomfortable'. The need to resign ourselves to the discomfort of the inevitable destruction is critical to this discourse. As one expert explained in a 2014 seminar at the Asian Development Bank office in Dhaka, 'accepting failure' of adaptation, instead of denying it, is critical to forging strategies for what to do in response to climate change. Another expert, working on a climate change project for one UN agency, told me more specifically (referring to plans to support the growth of shrimp aquaculture instead of adaptation of water management to support rice farming systems), 'we are not going to make things better with [river] dredging or water management, but there is definitely going to be shrimp farming'. By rejecting the possibility of strategies for effective mitigation of contemporary water management challenges, the practitioner effectively forecloses the possibility of an agricultural future for the region. This paradox of anticipatory ruination suggests that despite our *lack* of knowledge about what will happen, ruination is inevitable; moreover, despite the tremendous energies being committed to devising solutions, it appears already known that the only possible solutions will not pre-empt inevitable ruination of landscapes and communities.

Development planners and practitioners respond to this language of inevitable ruination with claims to the necessity of their own authority in devising responses. 'We need to create a vision for the future', explained one practitioner, precisely because of the need for uncomfortable solutions. Another adaptation expert explained to me that in this context, 'no one can know what they'll be doing in three or ten years down the road, including us, so as researchers and professionals, it's our job to figure out what that future will look like and then to introduce people to it so they can begin to adapt'. This speaks to a particular understanding of governance through knowledge production. Specifically, that the role of experts is first to imagine (in the absence of knowledge) what the future could or *should* look like, then to introduce it, and then to create it.

The normative dimensions of this discourse are most commonly articulated as an afterthought, while the difference between what is *possible* and what has been deemed *preferable* remains ambiguous. A program manager at one UN agency explained to me his project's support of shrimp aquaculture programs (and thus inundation of agricultural lands with saline water): 'in those areas, it's probably better to adapt to water. Water will come. It's probably much more beneficial than agriculture'. In this case, it remains murky whether the practitioner's assessment of the superiority of aquaculture to agriculture is due to its suitability to the ecological context, or some other normative claim about the relative economic benefits of shrimp vs. rice. What exactly is driving the change ('water will come') is also left unspecified. The active nature of the decision to bring water onto the land is obscured in such discourses. Similarly, a practitioner with one German-supported development agency objected when I asked him a question about why his program supported shrimp instead of rice production, by explaining 'no, shrimp is the major export earner, so we must support it'. But he then continued, 'in 10 years, who knows what will happen? We have no idea'. In both cases, the practitioner's own assessment of the economic importance of shrimp nullifies the question of the kinds of production systems that are and will be possible in the region in place of normative assessments of the production system that is most desirable. The role of programs like this in shaping the answer to this question of what will happen in 10 years, is also obscured.

Thus, whether the inevitability of ruination is due to the imminence of climate change or, alternately, a particular analysis of economic imperatives, remains an unlikely and unnecessary question. The anticipated future of the region is the same in either case. Development practitioners regularly refer to Bangladesh as a 'model', 'leader', and 'pioneer' in adaptation, in the sense that the very act of imagining the ruination of this region renders Bangladesh a test site for various modes of adaptation that can be applied in other settings. It also creates possibilities for the transformation and ruination as forms of adaptation in themselves (see Zee, this volume).

Shrimp

The expansion of shrimp aquaculture in Khulna plays a pivotal role in the region's anticipatory ruination, shaping both its context and realization. The effects of climate change, shrimp production, and the complex histories of dispossession tied up in development in this region since the colonial period, are all part of the interconnected dynamics of spatial governance at this frontier.

Prior to the shrimp boom, farmers in Khulna primarily grew one or two rice crops a year, and often a third 'winter crop' more amenable to the elevated salinity levels of the rivers in the brackish coastal region, such as watermelon and sesame. This production involved the labour of smallholders, sharecroppers, as well as landless labourers, the latter of which made up a majority of the population, owing to an extensive system of subinfeudation with roots in the colonial period (Von Schendel, 1982; Datta, 1998). The insecurity of land tenure resulting from this highly unequal distribution of land rights facilitated the dispossession that accelerated the shrimp boom (Adnan, 2013). It is the ruination of this agrarian political economy that has had the most profound influence on the transformation of production and social reproduction for Khulna's rural inhabitants.

Shrimp aquaculture has been implicated in a variety of different modes of social and ecological destruction in Khulna since the 1980s. Major concerns include: ecological destruction through chemical effluents from ponds, soil salination, severe reductions in aquatic species diversity due to bycatch through wild capture of shrimp larvae, mangrove deforestation for shrimp ponds in brackish water habitats, and water logging of soils due to the blocking of canals for shrimp enclosures (Ahmed and Troell, 2010). For local activists, these environmental issues are embedded in deeper social concerns, namely the dispossession from land and labour opportunities in the transition from rice to shrimp (Halim, 2004; Datta, 2006; Guhathakurta, 2008; Sur, 2010; Paprocki and Cons, 2014a). These dispossessions are inherent to the transition, embedded in a trans-formation in production relations that entails higher concentrations of land tenure and ownership and less intensive labour requirements (Belton, 2016). These impacts have been felt most acutely by women as well as the majority landless populations who have historically worked as sharecroppers and agricultural day labourers in the region (Halim, 2004; Datta, 2006; Guhathakurta, 2008; Paprocki and Cons, 2014b). For these local activists, resistance to anticipatory ruination entails an insistence on the possibility of continued agricultural production in the coastal region. Communities have mobilized both to continue farming rice as well as to return to rice cultivation after shrimp production had already begun. By insisting on rice agriculture, and the broader dynamics of production and social reproduction that accompany it, as a real alternative to shrimp, these communities contest the teleology of ruina-tion on this frontier.

Two particular modes of dispossession are most salient to under-standing these social dynamics. The first is the land grabbing that facili-tated the early waves of the shrimp boom starting in the 1980s. These enclosures were usually carried out by urban elites, often through the use of violence against local communities. Though these dramatic forms of violent land grabbing have declined in recent years, the insecurity and inequality of land tenure they escalated have not been ameliorated. Today, while violence continues to be used against those opposed to shrimp, land grabbing often takes place through various forms of judicial harassment and through the surreptitious inundation of agricultural lands with saltwater.[4] The second, and more far-reaching, mode of dis-possession is the reduction of labour opportunities in shrimp as opposed to rice production. Community members estimate that shrimp produc-tion requires somewhere between 1 and 10% of the amount of labour of rice farming. As shrimp ponds take over land from rice and other crop production, the people who used to depend on agricultural work to sur-vive and feed their families increasingly find themselves without a place in the rural economy. As the elderly proprietor of a tea stall commented to me, 'shrimp has destroyed all of the farmers', invoking the Bengali word *dhongsho*, meaning literally 'destroy', 'kill', 'waste', or even 'ruin'. These dispossessed farmers migrate out of their villages to find work – to Khulna city, Dhaka, and often Kolkata, in the neighbouring Indian state of West Bengal. Thus, the expansion of commercial shrimp cultivation has a significant impact on the transformation of labour relations throughout the region, as well as the survival of its inhabitants.

This migration out of the coastal zone reflects a process of depeasantisation (Araghi, 2009) that facilitates the intensification of resource extraction from the region. A landscape that once produced largely for local consumption is thus transformed to one that produces for export markets. In that sense, the shrimp boom has turned the region into a resource frontier (Barney, 2009). Yet, new discourses surrounding climate change in Bangladesh are re-shaping this frontier. While rural out-migration from Khulna is widely recognized (Guhathakurta, 2011), it is commonly framed as climate migration (Siddiqui, 2003; Norwegian Refugee Council, 2015). This framing not only obscures the dynamics of agrarian change at the heart of the transformation, it also facilitates further dispossession through the production of alternative landscapes in response to what is seen as inevitable ruination.

Sensitivity

The production of knowledge and imagined futures is important to the process of anticipatory ruination, but this construction of Khulna's ruination is fragile and incomplete. Mobilization of knowledge about the region and its ongoing changes, as well as the production of uncertainty, plays a critical role in shaping the space now and in the future.[5] The climate frontier is shaped by the constant production of knowledge about its spatial dynamics and integrity, paired with a multitude of silences and erasures, a studied suppression of particular histories and political dynamics that have shaped the space and understandings of it. The sensitivity of these knowledge formations plays an important role in the governance of the climate frontier. The lens of sensitive space, as Cons describes it, illuminates the 'anxieties, uncertainties, and ambiguities' at the heart of the production, experience, and governance of frontier spaces (Cons, 2016:25). Attending to the production of this sensitivity thus facilitates a better understanding of the governance of frontier assemblages more broadly.

Throughout my research, this sensitivity and uncertainty about the region and its futures was invoked in a variety of conversations about the drivers of change in Khulna and how they are understood. Specifically, the role of shrimp aquaculture in social and ecological degradation, concerns about land grabbing for aquaculture driving landscape transformation, and doubts about dominant narratives of the role of climate change, were all regularly deemed by donors, development practitioners, and government officials, to be off-limits topics of conversation, particularly for public discussion. The ambiguity of drivers of change, and maintaining the primacy of climate change as the principal explanation of ecological transformation, facilitates the spatial governance of the frontier. Thus, when I asked questions that called this narrative into question in the offices of these experts, they would frequently respond not by refuting, but by explaining that such topics were off-limits due to the 'sensitive' nature of the subject.

In five different instances during my research, respondents alluded to official documents examining research concerning socio-ecological and climatic changes in Khulna, but declined my requests to read the reports, citing concerns about their 'sensitivity'. For example, officials at one UN agency mentioned but declined to share a study on sea level rise, and researchers at one local institute declined to share a report on research they conducted on drivers of out-migration from Khulna (that seemed to suggest that climate change may not be the primary driver, counter to official narratives). In another instance, in September 2014, I visited a local government office in Khulna to inquire about the response to a petition submitted several weeks prior by local farmers. The petition, comprising well over 100 pages of signatures, protested against the inundation of farmland with saline water by would-be shrimp farmers in a rural community some 30 km from the city. The petition, according to local activists who had led the effort and presented it at this office, attributed soil salinity and degradation to the deliberate cutting of embankments to allow inundation, not to natural dynamics of erosion or saline ingress (as is often claimed in climate change narratives). The local official received me at his office, and after approximately two hours of obligatory tea drinking and niceties, finally sent an attendant to retrieve the petition with accompanying notes on the government officer's judgement, per my request. He delicately untied the strip of fabric that held the file together, placing it front of me, and after flipping through the pages of signatures, I began reading the official letter clipped on the top of the stack. Sensing my interest, the official suddenly closed the file and pulled it back from me, cursorily explaining that he could not let me look at it anymore because it contained 'government secrets'. His invocation of the sensitivity of this incident and the documents pertaining to it indicate the importance of excising the contingencies and contestations embedded in the record of ecological changes confronting the region.

Indeed, as Tsing examines in Indonesia, a frontier story requires a space without a history, empty of people (2000:131). Forging a frontier involves the erasure of these histories and the social processes that have composed them. The transformation of Khulna into a climate frontier

thus necessitates that silence govern narratives about the ongoing production of ecological crisis through the incursion of shrimp aquaculture.
The silence also necessarily extends to the dispossession of farmers and
rural labourers through the transition away from rice, and the social
movements these people organize to draw attention to the causes and
consequences of these changes. None of these narratives has a place in
the construction of the climate frontier. The deployment of categories of
secrecy, sensitivity, and erasure operates at the frontier to conceal how
these zones of ruination are actively produced through particular configurations of knowledge and imagination. Anticipation of ruination
through climate change at this frontier entails the erasure of other histories of ruination, and discourses that highlight the politics of ongoing
dynamics of ruination in the present beyond climate change.

Conclusion

The anticipatory ruination of Khulna through the expansion of shrimp
aquaculture follows the same contours of epistemic intervention as those
promoted globally through development planning in response to climate
change. In Khulna in particular, this logic operates as a progression,
masking the histories of each of its elements: uncertain future ruination
becomes an explanation for the inevitable unviability of agriculture;
shrimp aquaculture is proposed as an alternative; experts conclude
that it is preferable, anyway, and should be embraced immediately as an
opportunity. This entails a vision of what an ideal future is for communities in Khulna, divorced from those communities' own visions of what
is possible or desirable. Climate change becomes a post-facto justification
for a process of ruination through the planning of particular development
interventions both historically and into the future. Anticipatory ruination
produces frontiers materially in particular places while operating as a
broader epistemic project without geographical boundaries. This assemblage of epistemic and material dynamics thus both facilitates resource
extraction and forges new frontiers of production and accumulation.

Yet, frontiers, despite the discourses that accompany and produce
them, are not linear. Examining the dynamics of anticipatory ruination
in Khulna sheds lights on the stakes in succumbing to the teleologies of
frontiers in their material and epistemic dimensions. By understanding
the production of these frontiers, we are better able to understand not
only the dynamics of dispossession, but also the alternative futures that
are obscured in Khulna and elsewhere. The efforts of local communities
to highlight these dynamics of ruination help us to understand both their
antecedents, as well as to suggest the possibilities of alternative futures.

Notes

1 See Harris, G. (2014). 'Borrowed Time on Disappearing Land: Facing Rising Seas, Bangladesh Confronts the Consequences of Climate Change'. *The New York Times* (March 28).
2 These dynamics are embedded in a larger dynamic of governance that I have elsewhere described as an *adaptation regime*, a socially and historically specific configuration of power that governs the landscape of possible intervention in the face of climate change (Paprocki, 2018).
3 See 'Games for a New Climate'. (2012). Vimeo. https://vimeo.com/43127251.
4 Would-be shrimp producers or their agents often cut down protective river embankments at night.
5 I explore the politics of uncertainty in greater detail elsewhere (Paprocki, 2017).

2

Subsurface Workings
How the Underground Becomes a Frontier

Gökçe Günel

Preparations

On 4 December 2010, the United Nations Climate Change Conference (COP 16) in Cancun agreed upon the inclusion of carbon dioxide capture and storage in geological formations (CCS) as an eligible option for mitigating greenhouse gas emissions. CCS operates by procuring carbon dioxide from localized sources of pollution, such as power plants; carrying it in solid, liquid, or gas form to storage sites; and injecting it into the subsurface. Issues such as the selection of storage sites, monitoring plans for the leakage and seepage of carbon dioxide, and the transboundary effects of gas injection have to be actively addressed in planning for CCS projects. In further negotiating the means through which CCS would be adopted as a mitigation strategy, interested parties were invited to submit modalities and procedures guidelines to the United Nations Framework Convention on Climate Change (UNFCCC) that would address and resolve such controversies.[1]

Upon the declaration of the Cancun CCS decision, some policy consultants at Abu Dhabi's flagship renewable energy and clean technology company Masdar admitted that it was a surprise. This could be a turning point for an oil-exporting country like the UAE, they excitedly argued, as it would avail future options for low-carbon oil production and usage at the same time it would provide opportunity to earn carbon credits. By injecting carbon dioxide into fields, using them as storage sites, and forcing oil out, oil producers could extend the lifespan of oilfields.

Frontier Assemblages: The Emergent Politics of Resource Frontiers in Asia,
First Edition. Edited by Jason Cons and Michael Eilenberg.

Besides, it was a perfect occasion for the UAE to publicize its commit-ment to climate change mitigation goals, improving its image in the international policy sphere; the eventual inclusion of CCS as a climate change mitigation strategy could lead to a major diplomatic success, contributing to the UAE's transformation into a real leader in the international world. Therefore, when it was announced that interested parties could submit modalities and procedures guidelines regarding how carbon capture and storage projects should be initiated, maintained, and monitored, the consultants at Masdar immediately began working on a document, cooperating with other stakeholders: Abu Dhabi National Oil Co. (ADNOC), Abu Dhabi Co. for Onshore Oil Operations (ADCO), and the Directorate of Energy and Climate Change (DECC) at the Ministry of Foreign Affairs. The document had to be submitted to the UNFCCC by 21 February 2011.

This chapter tracks some aspects of the production of this CCS policy submission document, and draws on fieldwork with consultants at Masdar as well as with representatives from ADNOC, ADCO, the DECC, and the UNFCCC. In doing so, it takes the CCS submission doc-ument as a key moment in the forging of the subsurface as a new, imag-inative, frontier (see also Woodworth and McDuie-Ra, this volume). Similar to many others at Masdar, these environmental consultants had come to the UAE from all over the world for professional purposes and assisted in the production of national and international climate change policy, serving as key players in the climate debate. Beyond the UNFCCC, they pursued opportunities at intergovernmental institutions such as the International Renewable Energy Agency (IRENA) and at consulting companies, namely, the 'Big Four': Ernst and Young, Deloitte, KPMG, and PwC. They attended climate change summits, followed the debates related to various aspects of climate change governance, drafted reports for internal use and for the development of low-carbon technologies in other countries, and contributed to the policy production and imple-mentation landscape at the intergovernmental level. Internationally experienced, they often changed jobs, or moved between organizations, leaving the UAE for positions elsewhere within several years of their arrival. By drawing on how these consultants discussed the CCS docu-ment and investigating the language and perspectives they adopted, this chapter examines the work through which the underground became reconstituted as a frontier for burying carbon dioxide.[2]

By following the production of the UAE's modalities and procedures guidelines submission, I demonstrate that Masdar's investment in new technological developments was a fundamentally imaginative venture, ridden with risks and uncertainties – one that was central to the assem-bling of the subsurface as a frontier. In this context, Masdar constituted one among many dispersed nodes where the frontiers were assembled specifically to manage speculative and risky futures in a warming world. As such, debates over carbon capture and storage explored in this chapter highlight how resource frontiers in Asia are sites that are simultaneously made in place and bound up in broader, at times seemingly remote, transnational debates.

The environmental consultants and engineers in Abu Dhabi created 'degrees of acceptable risk' as they attempted to extend the status quo to an uncertain space and time through new technology. In negotiating the different ways of imagining earth – either as a temporary sociopolitical space or as a solid geology – they strove to come up with a shared sensibility regarding potential future risks and uncertainties. Formulated in this chapter as two main discussions, on 'particular and global levels' and 'unknowability of the subsurface', the tensions between different actors illustrate the significance of environmental consultants as mediators of risk management and also point to how 'degrees of acceptable risk' became crucial in producing new frontier spaces and climate change mitigation methods that might align with the fossil fuel economy.

Risks and Uncertainties

'We are working at least three kilometers underground!' Marwan, an Algerian consultant at Masdar, exclaimed, attempting to showcase the various risks and uncertainties of CCS during an interview in early 2011, 'If you don't have full understanding of what's underground, you cannot predict all the risks, and these risks will create problems in the future'. Marwan and I had met in the fall of 2010 while participating in meetings on the submission document, and saw each other often in the Masdar office in the years 2010 and 2011. He had studied chemical engineering as an undergraduate, and later received a master's degree in environmental science in France. After working with a large state-owned oil and gas company in North Africa for 15 years, he had accepted a project manager position at Masdar and moved to Abu Dhabi. He was focusing on conducting technical assessments and feasibility studies regarding domestic and international climate change policy proposals; in some ways, this was an extension of his previous work in producing environmental regulations for oil fields. He argued that his experiences with the oil industry gave him a rigorous perspective with regards to environmental challenges.

When I asked him how he would define the consultants' role in the implementation of CCS as a climate change mitigation strategy, Marwan continued: 'The whole exercise is a risk assessment for future risks.

We need to define the mitigation action and to reduce the risk. But the risk can be managed. All we have to do is to define who will be in charge of the risk, and mitigating the risk, and taking care of the accidents'. After referring to the Fukushima disaster in Japan, where following a major earthquake, a tsunami disabled the power supply and cooling of three Fukushima Daiichi nuclear reactors, causing a major crisis on 11 March 2011, and which at the time of our conversation shook the energy industry considerably, he underscored: 'There is always error. We just need to be prepared.' According to Marwan, the designers and operators of Fukushima nuclear power plant imagined that they knew everything about the plant, but they could not guess that a tsunami would hit and destroy its cooling facilities. What resulted was the second worst nuclear power plant accident in the world, ranked after the 1986 Chernobyl nuclear disaster. 'They had models for everything', he reiterated, 'but in reality things are different'. He framed the production of the modalities and procedures guideline submission as a process of manufacturing a set of directives to be used before and after an inevitable accident or disaster befalls a CCS project. Marwan's position was that through a surgical mapping and management of the risks and uncertainties, interested parties could implement a new mitigation technique at a planetary scale while regulating its possible harmful consequences. Marwan agreed that accidents were inevitable. Still he advocated that a strategy of mapping and managing of risks and uncertainties would help open the underground for re-engineering.

The multiple risks and uncertainties regarding CCS technology and policy, often referred to as merely technical problems that awaited resolving, may be considered metonymical expressions of a much larger project that was at play during the preparations of the modalities and procedures submission. In Abu Dhabi, environmental consultants, engineers, and scientists continuously tinkered with a possible redefinition of climate change as a commercially viable business. After all, they were responsible for facilitating business models to compete in the global fossil fuel economy in ways that might contribute to resolving social and environmental challenges. According to many of them, climate change had to be redefined in commercial terms to constitute a plausible topic of interest.

In producing a commercially viable climate change mitigation tool for the fossil fuel economy, the participants also disclosed their personal rifts, and tried to make sense of potential contradictions and inconsistencies. 'You must know first thing that any project should make money', urged Anand, an Indian environmental consultant at Masdar who was also involved in drafting the submission document, during an interview in his office. 'No one will do something eco-friendly and have loss.

Profits may be a little lesser but loss cannot be tolerated.' Anand held degrees in mechanical engineering and energy management, and had over 20 years of experience in climate change and sustainability services, having consulted for both private companies and governments. For most of his career, he had worked with one of the Big Four companies in India. During his time at Masdar, he focused on the development of renewable energy and climate change policy for the UAE, representing the country at UNFCCC negotiations, and strategized on the implementation of low-carbon technologies. In remarking that 'profits may be a little lesser but loss cannot be tolerated', he insisted on the impossibility of discussing environmental problems independently of commercial valuation.

During a long taxi ride from Abu Dhabi proper to Masdar, I asked Ben, a senior consultant to the DECC at the Ministry of Foreign Affairs, about the possible motivations for working within the emergent green economy (see Anderson, this volume). He responded, 'The environment is after all a sexy part of the economy.' Ben was working on the development and implementation of climate change strategy for the UAE by building policy, training diplomats, and attending negotiations. He was confident that green businesses were going to expand. He had been involved with the renewable energy and clean technology sectors for almost 20 years, but unlike Marwan and Anand, he focused his work on think tanks and nongovernmental organizations (NGOs). Before moving to Abu Dhabi, he had held a prestigious executive position in one of these institutions, and regularly travelled to climate change conferences. Reflecting on his experience working in multiple countries, Ben added that 'not many of the environmental consultants would self-identify as environmentalists'. Still, the development of the modalities and procedures submission served, for many of those involved, as an exercise in bringing together moral domains that may not be immediately compatible with each other – and indeed may be incompatible.

This dilemma was exacerbated because of the UAE's potential as an oil-producing country. When implementing CCS technologies, oil-producing countries are able to use maturing oilfields as storage locations for the carbon dioxide that they obtain from industrial compounds, as these oilfields may be considered naturally sealed reserves. However, injecting gas into oil reservoirs leads to increased oil production as well, a process commonly known in the industry as enhanced oil recovery (EOR). By injecting carbon dioxide into ageing fields and pumping oil out, oil producers may increase the lifetime of the fields by up to 30% while freeing up the natural gas more commonly used in such processes. The inclusion of CCS as an eligible technology for decreasing carbon emissions then becomes a perverse incentive for further oil production;

the entities that earn most carbon credits from CCS activities in turn become oil-producing countries.

For many oil producers, including the UAE, it was important that CCS-EOR become recognized as a climate change mitigation strategy, especially because of its immediate applicability and eventual financial gains. During a meeting with representatives from the DECC at the Ministry of Foreign Affairs, some consultants suggested that the UAE's oilfields are still young, and would only require EOR in 30–40 years time. Ben, in advance of his participation in the next UN climate summit, asked how far he should push for EOR. 'Yes, EOR is significant for us', one consultant responded, 'but we would like CCS to be approved as a climate change mitigation technique even if EOR is not approved'. Another consultant continued, 'We need policy to pay for the environmental premium that we invest in this new technology, we are not looking for a cash cow here.' They discussed a potential workshop, wherein experts from the Association of Small Island States (AOSIS) – at the time a major opponent to CCS as a climate change mitigation strategy – would be invited to the UAE, or maybe to Norway, to be shown how CCS functions on the ground. 'Bring them, and convince them here', one DECC representative ordered. In the meantime, Masdar was publicizing plans to capture 800,000 tons of carbon dioxide from Emirates Steel's plant in the Mussafah industrial zone, and to transport it through a 311-mile pipeline to Abu Dhabi's oilfields by the end of 2012. During a lunch break, Ravi, who had been working as a consultant for almost seven years, confessed to me, 'Personally, I am against CCS-EOR. But as a consultant, I will do everything to make it approved.'

Emergent technological projects open up new moral conundrums for the individuals that research, develop, or implement them (Fischer, 2003). Ravi attempted to resolve his moral dilemma by constructing multiple, at times divergent, perspectives, thereby preventing his personal goals or values from counteracting with his professional commitments. An Indian engineer in his early 1930s, he had held energy management and consulting positions in India and the UAE, and would soon be moving to Europe for another opportunity. He did not think that EOR projects should receive carbon credits or any further encouragement from policy-making institutions, as they already generated additional amounts of fossil fuels to thereby profit. But in the end, as he saw it, climate change policy was too extensive of an issue for him to effectively influence – his presence or absence in the debates would not change the outcome, they could easily find someone else to do his job. The questions of scale overwhelmed Ravi, and made him feel small and unimportant. The scale of the problem also gave him a particular freedom, where he could act without worrying too much about his own beliefs or actions, submitting to a larger network fully capable of acting in his absence. 'Climate change is a business opportunity', he concluded. 'We should try to come to terms with that.'

Where do Ravi's moral conundrums fit when conceptualizing the production of the underground as a frontier? As he suspected, Ravi's dilemma was assigned a place neither in the submission document nor in the larger bureaucracy that the document represented. Instead, Ravi concentrated on a technical and conceptual debate, taking place between consultants and engineers on whether the subsurface should be conceived as a black box, and managed to overlook his moral and ethical predicaments. His investment in completing the submission process in a timely and successful manner not only triggered but also stifled the larger questions regarding the potential social or political meanings of creating this policy in the first place.

Subsurface is (not) a Black Box

Increased oil production, and therefore consumption, is not the only reason why CCS activities are considered controversial as climate change mitigation strategies. Parties critical of CCS projects point out that besides issues of site feasibility, high operational costs, future safety, and unresolved legal liability that make CCS projects challenging to initiate, implement, and operate, CCS may incur a crowding-out effect, leading investment away from other climate change mitigation strategies, such as renewable energy or energy efficiency projects.

In considering such existing risks and uncertainties, policy consultants at Masdar would pose a wide range of questions to one another. 'What if we pump carbon dioxide into the ground here and it comes out of Iran?' one of them asked during a preparatory in-house meeting, emphasizing the urgency of international legal protocols for resolving liability issues. 'What if there is seismic activity due to CCS? Who will be responsible for it in 10,000 years time? Who will be responsible for it if nation-states disappear?' another consultant brought up, pointing to the deep futurity of current geological engineering projects. They discussed how models for storing nuclear waste could help in planning for carbon dioxide storage. They also spoke about producing new insurance mechanisms to involve new safety funds. Consistently, the indeterminate spatial and temporal boundaries of CCS made it challenging for consultants to conceptualize and attend to the safety and liability issues associated with upcoming projects, and prompted them to mobilize multiple temporal and spatial scales at once. Despite bringing forth challenges, these conversations allowed the consultants to imagine and assemble the subsurface as a frontier.

In studying the negotiations for an international legal instrument at a global United Nations (UN) conference Annelise Riles has concluded, 'The 20th-century problem of international institutions has been one of how to grasp [the global, national, and regional] levels, at one and the same time, how to bring them into a single encompassing view' (Riles, 2000: 91). The 'levels' problem that Riles identifies did not emerge as a question for the policy consultants at Masdar initially, either during preparatory in-house meetings or in meetings with the Ministry of Foreign Affairs. As time passed, however, the intrinsic characteristics of CCS technologies introduced shifting levels into their debates. The consultants at Masdar addressed questions not only about nation-states, but also about their potential disappearance. They highlighted the arbitrariness of borders, in addition to their transitory nature. They showed that the present, as manageable as it may seem, was not strong enough of a temporal unit to work with. In this sense, the document preparation process disclosed and buttressed the precariousness of the categories that international institutions simultaneously relied upon and reproduced, consistently demanding an imagination of alternative futures and alternative geographies.

And yet these alternative future scenarios had to be 'translated' into the document as it was being composed – formulized according to the thinking that the consultants could presently afford. The consultants had spoken about the possible disappearance of nation-states during their discussions, but when they put down their liability scheme on paper, it did not directly address the possibility of a world without nation-states. Instead, the submission document explained that project proponents would remain liable during the short lifetime of the CCS project, but since the lifetime of underground carbon dioxide was much longer than the lifetime of a CCS project, project proponents would later transfer the liability 'to an authorized body designated by the host country after the end of their short term liability period'.[3] The financial provision, which the project proponent would provide, had to cover this long-term liability period, and had to 'be based on a long-term probabilistic risk assessment, to be approved by the local authorities as per agreed international rules'.[4] In short, the submission document pushed the consultants to think about wide temporal and spatial scales and simultaneously locked them within the boundaries of the present.

In another instance, the consultants focused specifically on transboundary projects, and engaged in brainstorming about potential liability schemes, mostly by thinking through bilateral agreement methodologies. They wondered how the situation could be managed if several storage projects were conducted in the same reservoir, or if the reservoir stretched beyond more than one area of jurisdiction. 'It may make sense to initially limit to single projects', they suggested, meaning that a storage site would have to be entirely located within the territory of a single nation-state. Later, they spelled out the various impossibilities of their protocol. Someone incisively asked, 'How do you *know* whether a project will have transboundary impact?' Finally, when putting their thoughts on paper, the consultants simply concluded, 'We should have a requirement for upfront liability agreements'. The brainstorming sessions demonstrated that the boundaries and temporalities of the emergent frontier were indeterminate.

In contrast, the reservoir engineers, geologists, and geophysicists who participated in mid-stage meetings proposed solutions for these problems right away. A particularly telling exchange occurred at a meeting in early February 2011 between Anand, who had been a significant voice at all the early stage meetings, and Shahab, an Iranian executive at Abu Dhabi Co. for Onshore Oil (ADCO), who began to participate in the drafting process only after the initial theoretical parameters had been outlined by the consultants during in-house discussions at Masdar. 'I don't like documents which are only buzzwords, that don't have any meat', declared Shahab upon reading the submission draft. Shahab had earned a PhD in oil recovery and reservoir management from a US university in the mid-1980s, and occasionally taught classes and advised theses on these topics. 'We should give more details about the local setting, which we know very well, only then can we produce satisfactory risk matrices', he continued. But Anand, speaking on behalf of the Masdar's team of consultants, quickly rejoined that they were striving for 'a floating language' rather than 'a fixed one'. The floating language that he entailed would be elastic enough to contain global histories and geographies while retaining its meaning at a much smaller scale. Its vagueness would be its strength.

Another policy consultant tried to explain that the document had to be generic, and applicable to all countries that are part of the Kyoto Protocol, an international environmental treaty that set carbon emissions limits for all its signatories, and that went into force in 2005. But Shahab protested, 'but then how are we going to account for the differences between carbonate oil wells in the UAE and sandstone oil wells in Europe?[5] How will we account for the difference? Where is there anything that shows this will be stringent or robust?' Shahab insisted that the submission include more details. In the meantime, Anand encouraged him to study the issues from a more macro perspective, and grew exhausted. 'This is all about negotiation – what is our approach, what are our tactics? We need to state what we want without being apologetic. Descriptive words will make a reasonable submission. Make references to all tools, but make no commitments', he summarized. 'Think that you are sitting in the UN and developing guidelines for the globe.' While the consultants took the shifts between the local, the global, and the regional for granted, it proved very difficult for them to communicate these necessary 'levels' to their ADCO and ADNOC counterparts. 'High level phrases, and not details – that's what you want?' Shahab confirmed in a rather resentful way before he left.

Extending his arguments about the inevitability of errors and accidents, Marwan, who had participated in the preparatory meetings, noted to me in May 2011, 'The subsurface is a black box'. He then drew a graph (Figure 2.1) in my notebook to communicate the impossibility of representing or knowing earth's geology. The graph, with unidentified x or y-axes, indicated that in attempting to represent the underground the steady 'model' and the wobbly 'real' intersected only briefly, in this case at four specific points. He explained that the representation efforts that took the shape of models were loaded with assumptions, and therefore unable to accurately pinpoint the real. Marwan's framing of the subsurface as blackbox highlighted the imaginative process of assembling the subsurface as frontier of carbon storage. A belief in this ultimate unrepresentability or unknowability, conveniently contained within this drawing, had made it easier for the consultants to switch levels and contexts between, let's say, carbonate wells in the UAE and sandstone wells in Europe, without fully disputing the physical qualities that they manipulated. As such, for the consultants, uncertainty and risk remained constitutive of policy-making efforts on CCS. They suggested that they needed to act like attorneys, and defend the submission as if it is a legal case.

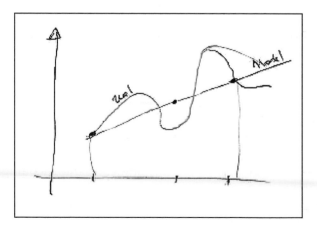

Figure 2.1   Real vs. model graph drawn by Marwan, May 2011. Photo: Gökçe Günel.

As Timothy Luke states, moral dilemmas and obligations are ascribed clearly demarcated niches within technical, managerial, and organizational sets of problems and solutions – a mode of management that defines and reflects how environmental challenges should be handled in the present (Luke, 1995). In his examination of environmental politics in Hong Kong, Timothy Choy also reports how 'an environmental consultant based in Hong Kong drew a provocative analogy to describe his role: We're just like lawyers, only with science. A client hires you, and you argue their case. But we use science rather than the law' (Choy, 2011: 84). At Masdar, the legal metaphor that the consultants used allowed them to legitimately distance themselves from this task, and enabled them to feel less responsible for any potential implication their claims or decisions may have. The unknowable variables raised in their brainstorming sessions (as well as moral conundrums, such as Ravi's) had mostly been left out of the submission document. In defending the document as if it was a legal case, the consultants knew and accepted that it included many unanswerable questions, but still the assembling of this particular frontier required them to produce 'degrees of acceptable risk' regarding future CCS projects. The strategies of risk management constituted their main tools for manufacturing such effect.

After the document draft was completed and submitted to the UNFCCC in late February 2011, I visited Shahab and his team at the headquarters of ADCO, hoping to learn more about how they conceptualized the mechanics of the underground. A simulation expert, who worked with Shahab, invited me to his desk to show his work on dynamic models: 'It is not only oil that comes out of the reservoirs', he said, 'We are also digging for information about the subsurface.' He was seated across from a large computer monitor where he manipulated the data that had been collected since 1973, when the first large-scale drilling operations started in Abu Dhabi. He switched from one year to another, changed reservoir pressures, zoomed deeper, 7 km below ground. Finally, when he was convinced with the aesthetics of the resulting image – coloured in bright pinks and greens – he pressed print, and handed me the sheet (see Figure 2.2). Someone else in the office later commented, 'The subsurface is *not* a black box'.

Embellished with injectors and extractors that resemble spikes, this colourful model was an in-depth representation of pressure levels in the underground at a certain moment. Such a space-and-time-specific model is significant for pursuing geophysical studies, as Geoffrey Bowker notes in elaborating the practices of industrial geophysics during Schlumberger's initial years – before Schlumberger became the world's largest oilfield services company, operating in 85 countries. He states, 'There was never a perfect fit between figure and ground, and "general" results were

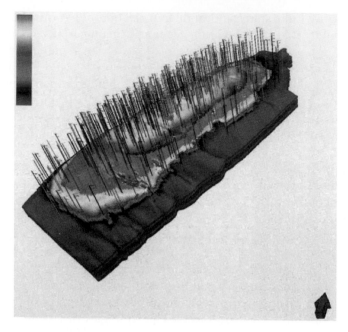

Figure 2.2   Subsurface simulation model, 2011. Photo: Gökçe Günel.

highly particular, rooted to a single site' (Bowker, 1994: 32–33). In this way, he claims that geophysics is similar to medicine: it is a science of the particular, deriving its data and treatment techniques from individual cases. This perception of geophysics may make it somewhat easier to understand the problems that Marwan or Anand suffered from in their relationship with the scientists and engineers at ADNOC and ADCO.

While the scientists and engineers believed they would be more successful in attending to individual cases, and pushed the consultants to include case-specific data in the submission document, the consultants refused such contributions, dictating a global frontier thinking. But the geophysical research methodologies did not allow for the floating language that the consultants demanded, thus triggering an inevitable disagreement between the two parties. As such, geophysics unveiled how 'the subsurface is *not* a black box' for particular instances, but when it tried to produce more generalizable models, it did fail, reflecting what Marwan called the 'real' in only a few instances. The discussion between Anand and Shahab had in fact exposed a pertinent disciplinary characteristic, which possibly made it challenging for the scientists to switch 'levels'. Accordingly, the discussions regarding the 'levels' and 'the subsurface as a black box' called attention to the difficulties of the transition from the particular to the general or vice versa, in different manners. The two knowledges utilized different means of abstraction in linking the particular to global levels. The subsurface emerged as frontier, in part, in the space between these tensions.

The reservoir engineers and geologists at ADCO claimed an ever-improving knowledge of the subsurface, and evidenced multiple research techniques. Static and dynamic models, in addition to seismic and sonar testing, were crucial in reproducing the underground. They did not have any predictive models at hand, but by relying on a combination of computer modelling and 'real-world' monitoring, they would excel at CCS technology. 'But it is all about the imagination at the end', Aref, an Emirati reservoir engineer at ADNOC, reflected. He explained that his company organized field trips to Jebel Hafeet (Figure 2.3), the UAE's best-known mountain – 4,098 ft tall at its peak – in order to expose the earth's geology to students and junior engineers. 'We explain that there are infinite layers, like the layers you see on the mountain, inside the earth', he continued. The students and junior engineers would observe the mountain, and extrapolate.

The imagination work that the engineers attempted, however, was rather different from the brainstorming sessions at Masdar. Aref stated that such a geological field trip could at times be unnecessary. To understand an oilfield, a student could simply observe a child sipping a drink by using a straw. While every oilfield had particular qualities, what

Figure 2.3 Jebel Hafeet, 2011. Photo: Gökçe Günel.

mattered most was to understand simple mechanics that could be applied in multiple contexts. The engineers at ADNOC and ADCO took these infrastructures for granted, in a way that the consultants at Masdar could not. Both parties sought applicability and transferability of conditions. But while the consultants addressed a possible transformation of sociopolitical contexts, albeit temporarily, the engineers and scientists relied on the seeming stability of the earth's crust, and its associated mechanics.

Having occupied various positions in the petroleum industry for more than a decade, Aref had worked numerous times with optimum field development plans, which are designed using subsurface static and dynamic models. After receiving his undergraduate degree in mechanical engineering in the UAE, he completed a master's in France, and specialized in well performance, reservoir simulation, and reservoir monitoring at other oil companies. A week after our meeting, over an email exchange, he suggested that these tools were still quite limited, but that they were progressing very fast. 'If we focus on the task of collecting all relevant data and building the right tools to model the global climate phenomena, we will be able to play a few global weather forecast scenarios assuming different levels of human activity. Only then, should we be able to seriously think about deliberate manipulation of planetary environment', he concluded. Perhaps the engineers did not believe that the planet was ready for policy-making regarding geological engineering projects such as CCS. An executive from ADCO asked, 'How could you generate policy without proof of concepts? We may be ignoring the key risks.'

The trouble, then, must have been about differing conceptions of risk and uncertainty. Many consultants believed that they had to complete the submission in a way that matched or even exceeded global expectations, regardless of what may lie ahead in terms of CCS or EOR research. On the other hand, the reservoir engineers and geologists attempted to localize the risks and uncertainties, and to produce as much quantitative feedback as they could, which would not be directly useful for other contexts.

How does one understand the relationship between the different ways of imagining earth – either as a temporary sociopolitical space, or as a solid geology – and the different ways of perceiving risk and uncertainty? In trying to create the modalities and procedures guidelines submission, the consultants knew and repeatedly admitted that they dwelled on risky and uncertain ground. Nevertheless, they tried to conjure a possibility of certainty, charted in vague, floating sentences, or uncertainty matrices. In a way, they utilized this document as a means of restraining their imaginations, or a mode of creating stable frontiers. Yet, the engineers and scientists who were present in the debates thought that their imaginations already gave way to a level of certainty, especially at a particular, case-by-case basis. As such, the charts were only a means of representing the stability that they had access to by observing the earth's dynamics – during a mountain excursion, or while watching a child sipping drinks.

Soon after his fevered discussion with the ADNOC and ADCO engineers and geologists in February 2011, Anand reiterated, 'You cannot reach a destination when you have no road.' As indeterminate as they were, the policy-making efforts were more than valid because they would eventually enable engineers and scientists to work towards the betterment of climate change mitigation technologies. They would serve as infrastructure, more than anything, and pay for the environmental premiums which governments and corporations were expected to invest. Anand could not understand what it was exactly that the scientists and engineers misconceived.

What did different conceptualizations of whether the underground could be 'black-boxed' mean for the production of the underground as a frontier? These technical and conceptual disagreements, which had to be resolved before the conclusion of the submission process, constituted an indisputable focal point, and allowed white-collar professionals, such as Ravi, to defer moral and ethical questions in reflecting on their work. These daily challenges, not only disclosed epistemological tensions between consultants and engineers, demonstrating that the two groups employed different means of abstraction for linking the global to the particular, but also comprised a top priority, pushing professionals to generate a shared sense of risk and uncertainty, and relegating other possible debates to the background.

Depicting Climate Futures

In March 2012, about a year after the submission of the guidelines, a German UNFCCC secretariat representative told me about the potential dangers of concentrated amounts of carbon dioxide as we had lunch in a Bonn restaurant. I was conducting fieldwork in Bonn to learn more about the ways in which UNFCCC secretariat members evaluated the CCS submissions they received from various countries, such as the UAE, and had been invited to an informal gathering with members of the CCS team. The German secretariat representative began by referring to an incident in Central Africa in 1986: Lake Nyos in Cameroon started emitting large levels of carbon dioxide, leading to the large-scale asphyxiation of humans and animals in the surrounding areas.[6] The carbon dioxide that leaked from the lake had been completely natural, and demonstrated how fatal such out-gassing could be. There were recorded instances of outgassing in Canada as well, where CCS experiments had taken place for some years. Canadian officials did not know to what degree the emissions resulted from injected carbon dioxide – their accounting had come to a dead end. The UNFCCC representative then pointed out that some states (*Länder*) in Germany were now banning potential CCS projects. A 2011 International Energy Agency (IEA) report entitled 'Carbon Capture and Storage: Legal and Regulatory Review' included the following assessment: 'CCS is still highly controversial in Germany .... Additional controversy has been generated by the inclusion of an "opt-out" clause in the draft act at the insistence of certain *Länder*, whereby states can designate areas as ineligible for CCS deployment, effectively vetoing CCS in those areas'.[7] New CCS regulations became effective in August 2012 in Germany, only per-mitting research, pilot and demonstration projects, and allowing for the possibility of restricting carbon dioxide storage to certain parts of the country.[8]

'Right now things are dismal', another UNFCCC representative briskly commented on climate change policy during an interview in his Bonn office. According to him, the main risk regarding CCS specifically and climate change generally was about commitment and willpower issues, where decision-makers did not always exert control to reduce emissions or employ the adequate mechanisms to mitigate climate change. 'The United States has its own issues to deal with, and the Republicans are hard to crack', he said. Otherwise, he believed that the UNFCCC had a clear roadmap to resolving uncertainty and risk. As we wrapped up our conversation, he told me, 'Solutions are available, and willpower will create clarity, and uncertainty will be resolved in that manner.' He was not playing environmentalist – he specified that he understood the profit-making needs of industries. 'I would not take a rustic action and propose a radical position saying close everything down', he concluded.

Like many other environmental consultants at Masdar, UNFCCC representatives endorsed an extension of the status quo, and partici-pated in the assembling of the underground as an imaginative frontier.[9] In pushing for climate change mitigation technologies like carbon capture and storage, they also acknowledged the uncertain conditions at hand. Once again, 'degrees of acceptable risk' emerged as a useful tool – involving the recognition that carbon dioxide was possibly dan-gerous, but, with effective expertise, its effects could be kept under control.

Notes

1 For a more extensive review, see Günel (2019).
2 For another examination of CCS technologies, see Günel (2016).
3 See submission document: 'Submission of views from the United Arab Emirates on addressing the issues referred to in paragraph 3 of FCCC/CMP/2010/L.10 in the modalities and procedures for the inclusion of carbon dioxide capture and storage (CCS) in geological formations as clean development mechanism project activities', p. 24
https://unfccc.int/files/methods/application/pdf/uae_submission_on_ccs_in_cdm_20110221.pdf.
4 Later in the negotiations this provision would continue to be controversial, as government representatives would be reluctant to take over liability for carbon dioxide.
5 The chemical behaviour and porosity and permeability levels of carbonate and sandstone rock formations are argued to be different from each other, which may require differing levels of risk and uncertainty as well as customized monitoring criteria. For more information on the differences between two types of reservoirs, see Ehrenberg and Nadeau (2005).
6 Elaine Shanklin writes: 'In the night of 21 August 1986 Lake Nyos exploded. The 'good' lake, as the locals called it, the most beautiful crater lake in Cameroon's North West Province, exploded and sent down to the valley beneath a deadly cloud of carbon dioxide that killed most of the living things it touched – 1,746 men, women and children, more than 3,000 cattle, plus countless numbers of sheep, goats, birds and insects. Little or no damage was done to plants, crops or inanimate property. Houses, market stalls, village ovens and motorcycles stood untouched, while their owners lay dead nearby' (Shanklin, 1988).
7 International Energy Agency (2011). 'Carbon Capture and Storage: Legal and Regulatory Review'. http://www.iea.org/Papers/2011/ccs_legal.pdf.
8 For a report on German CCS legislations, see: https://hub.globalccsinstitute.com/publications/dedicated-ccs-legislation-current-and-proposed/german-ccs-legislation.
9 After concluding fieldwork in the UAE in June 2011, I followed carbon capture and storage negotiations at the United Nations climate summits in Durban and Doha. For an analysis of policy-making in Durban, see: Günel (2012).

3

Groundwork in the Margins
Symbiotic Governance in a Chinese Dust-Shed

Jerry Zee

Dust's Edge

The rolling dunes of Inner Mongolia's western plateaus undulate wildly out of the side windows of the shiny Chinese-made jeep that Forestry Secretary Xiao has claimed for the week. From the smooth new government road piercing its way through the scrublands that buffer it against encroaching dunes, the landscape appears as a frenetic rhythm. 'This area, because of its proximity to the Helan Mountains', an inland range cutting rain shadows into the fickle clouds, 'receives much more water than deeper in (*litou*)', he explains, pointing toward the yellow streak of dunes. By 'deeper in', he means the open desert, the vast belly of the continent extending west from the province's western frontier, as though the jeep were skating on the rim of the earth. 'That is why we can seed here.'

The road, beset by spreading sand, marks Alxa as the frontier of a national meteorological crisis. The dust storms that rise off the sandy surfaces of the increasingly desertified Alxa Plateau, in the far west of China's Inner Mongolia Autonomous Region, have since the early 2000s earned the region the dubious distinction of 'cradle of dust storms', when two seasons of upwind drought across northern China registered as Beijing's two most catastrophic dust storm seasons since the declaration of New China in 1949. More than a thousand kilometres and a full day from Beijing as a dust storm flies, the remote reaches of sparsely populated Inner Mongolia appeared in dangerous proximity by the wind.

Frontier Assemblages: The Emergent Politics of Resource Frontiers in Asia, First Edition. Edited by Jason Cons and Michael Eilenberg. © 2019 John Wiley & Sons Ltd. Published 2019 by John Wiley & Sons Ltd.

Beijingers demanded a form of environmental protection that would disrupt the continuity of spreading sand into strange weather: protecting Beijing from the meteorological fallout of devastating upwind desertification, or, the protection of some environments from others.

Inner Mongolia, a region of China whose coal has powered the Chinese economy and unleashed a boomtown economy of rampant resource extraction (Woodworth, 2017), has in the same stretch of time emerged as a meteorological frontier for the developed cities of the Chinese coast. In 2000, after a catastrophic season of eight closely-spaced dust storms over Beijing in the course of two months (Stein, 2015: 320–321), then-Premier of the People's Republic of China Zhu Rongji left the capital, wending his way against the trajectory of a dust storm until he arrived at the Alxa Plateau. There, he announced that desertification was not simply a problem of local, if catastrophic, land degradation, but rather an atmospheric and political problem for the country as a whole. If the desert continued to advance, if the earth could not be held steady, he warned, 'Inevitably, China will have to relocate its capital.'[1] The announcement punctuated a broader administration reworking of China's interior as Beijing's dust-shed, a zone of social, economic, ecological, and technical experiments (Ong, 2006) installing themselves across the landscapes where a desert might become a storm, and a frontier might press into the centre. Supplemented technically through back trajectory analysis, remote sensing, and the proliferation of dust monitoring and data collection stations, desertification and dust storm politics rework China's interior as a new kind of margin, in which relations of centre and periphery are coordinated as points in the development of a problematic meteorological process.

Dust-Sheds and the Proximate Frontier

This chapter concerns itself with this doubling of a resource frontier into the headwaters of a meteorological catastrophe, and experiments in social and landscape governance that have aimed to hold the earth to the ground (Zee, 2017). It hangs in a reworking of frontier regions in China, historically sites of imperial encounter and later contentious ethnic politics (Rossabi, 2004), into key sites in a geography of meteorological vulnerabilities. This reworking is closely tracked to the dust storms that have become a mark of urban life in contemporary northern China as much as the vertiginous social and economic change that has accompanied decades of Reform and Opening and its attendant social and market experimentation (Anagnost, 1997).

As the editors of this volume suggest, frontier regions have often been framed as distant, lawless regions. It is also true that they are made imaginatively, materially, and governmentally as sites of political experimentation; their status as margins is central to their productivity as incubators of political possibility and mutation. For dust storm source regions, the distance of the frontier is tempered by a paradoxical proximity through airstreams that draw together farflung places as stations in a possible storm. Since the early 2000s, China's desertifying Inner Asian frontier has increasingly appeared through the vexed and simultaneous figures of problematic meteorological closeness on the one hand, and the furnace of China's economy through decades of coal boom. These contradictions and the political experiments that they necessitate recall Owen Lattimore's seminal reflections on China's Inner Asian frontiers as sites of complex and negotiated sovereignty (1940). Two decades of concerted central government attempt to manage the content of Beijing's air have reworked northern Chinese airstreams and upwind desertification into the administrative footprint of a political apparatus spilling over what appears, through remote sensing and on-the-ground dust monitoring sites, as the capital's dust-shed. The distant resource frontier thus appears as a startling proximity, describing a meteorological entanglement playing out over the several days and hundreds of miles that connects places like Alxa to Beijing in a dust storm.

As the Inner Mongolian fringe doubles into a meteorological proximity, frontier ecology and environments are remade in dust control campaigns as an open laboratory of social, economic, and ecological re-orderings (see also Choi and Woodworth, this issue). In particular, this chapter aims to think through the various ways that figurations of 'environment' in experiments that draw together market interventions and forestry strategies generate emergent notions of political and ecological subjectivity. 'Environment', I suggest, is emerging as a governmental template for the gathering, manipulation, and control of social, economic, ecological, and geophysical dynamics in China's meteorological frontiers. That is, it appears as a contemporary means of assembling the frontier as a matter of state organization of relations that cut across staid distinctions between human, technical, and natural domains.

Thinking land, air, and more-than-human life in relation to shifting demands for economic development is a way of approaching the slipshod implementation of what has been grandly hailed in Chinese officialdom as a coming 'socialist ecological civilization'.[2] I approach this question through the fortuitous symbiosis of two plant species, whose planting has offered a tentative rubric for imagining and implementing a reworking of socialist and postsocialist institutions and social formations into part of an environmental strategy for changing the physical characteristics of a windprone landscape. I argue that this symbiotic reworking of governance – a government modelling of social and geo-physical entanglement through botanical entanglement – deploys the semantic potentials of the notion of environment, in English as well as Chinese, to posit a political logic in which 'environment' is both an object and a tactic of governance.

As anthropologist Lisa Hoffman reminds, 'environment', *huanjing* in Chinese, as in English is polysemous. It refers to extra-human nature, but also, more abstractly, to an arrangement of things, a milieu in and out of which things can be made to happen, the set of 'conditions that make something possible' (Hoffman, 2009: 107). An environment is a formal principle, as well as an account of the event through its shifting infrastructural conditions. It is a set of conditions in which something is held, and it is open to rearrangement.

I explore how the management of a diverse set of human and more-than-human domains as part of an integrated yet internally heterono-mous environmental strategy introduces new modes of managing human behaviour and its environmental consequences. Put succinctly, the con-ditioning of economic environments – and thus the rethinking of an economy as an environment among others – has become a key strategy for conditioning ecological environments in Alxa. It does so without ref-erence to categories of belief or care, instead foregrounding naturalized notions of economic rationality as a general theory of human action. In this, I engage with and supplement important theses on environmentality, the creation of environmental subjectivities. I gesture toward frontier experiments that re-assemble economic and ecological convergences into environments for reshaping human behaviour as a site where we might consider state environmental programs that do not aim to create 'people who care about the environment' (Agrawal, 2005: 162) as much as they seek to continually coordinate human labour into desired ecolog-ical formations (Figure 3.1).

In what follows, I focus on one project of experimental social and landscape governance. Forestry quotas for the planting of *suosuo*, a sand-binding shrub and the key species for windbreak afforestation in Alxa, especially on sandy former pastures, has hinged on the promotion of the planting of a second root, *rou congrong*, that grows only by inter-rooting in the micro-environment created by established *suosuo*. *Rou congrong* is a potentially valuable sand ginseng, and local forestry offi-cials bet on creating markets for this second root as a way of enticing ex-pastoralists to grow *suosuo* as its symbiotic precondition. Out of this symbiosis, the entangling of economic and ecological formats give shape to modes of more-than-human environmental governance (Figure 3.2).

Figure 3.1 In a *suosuo* forest, inspecting *rou congrong* bodies allowed to seed for next year's crop. Photo: Jerry Zee.

Figure 3.2 *Rou congrong* supermarket. Photo: Jerry Zee.

Groundwork

March 2012: Alxa. I've tagged along with Mr. Li in his jeep, as he rides over the high dunes that were once his pasture. Mr. Li has gotten lost, and he stops multiple times to radio the leader of his hired planting team on a walkie-talkie. Here, as winter turns to spring, the frozen earth thaws into windblasted motion, spraying loudly against the metal doors of the car. Li assures me with gruff good humour that he's sure the forest is here somewhere. He pauses to get his bearings before he punches the ignition, sending the jeep barreling down the windward face of a high dune. When he sees the red flag and the faint silhouette of water trucks, he hits the gas gleefully. This corner of what was once pasture is now a field of sand drifts, and in the short springtime planting window, it is abuzz with activity. Workers are busy planting shrubs in rows at right angles to the wind. They wear bright scarves over their faces, protection from the dusts that whip around our boots and peel off the surface of the dunes.

I walk with Mr. Li as he inspects the workers, mostly ex-herder women hired as day labourers, moving in groups of three. They plant files of *suosuo* (English: saxoul, Latin: *Haloxylon ammodendron*) saplings in straight lines across the undulating topography of sand dunes, one sinking a shovel into the loose sand, a second carrying a bundle of seedlings, a third setting them upright in the ground. Moving slowly between these rows, men, also ex-herders, drive trucks hauling water pumped and purchased from a neighbour's deep well. They spray and tamp down wet sand around the scraggly root balls of the saplings.

From a tentative promontory on the windward edge of a dune, Mr. Li invites me to imagine this expanse of sand as a future forest of shrubs, dampening the hard wind into a pleasant breeze, and catching the wayward dust in the screen of its brambly foliage. *Suosuo* is a hardy shrub adapted to the environmental and climatic conditions of the desertifying Alxa Plateau. Quotas for its planting emphasize its status as a key species in the mega-forestry projects for re-engineering shifty desertified topographies into geophysical stabilities.

For a local government charged with the task of sand control, *suosuo* is a key species. Forestry officials value the plant's quick-growing roots as an effective 'biological method' (*shengwu fangfa*) of dune stabilization (Wang, 2009), an ersatz sand-binding and wind-breaking infrastructure on and below the shifting surface of dunes. *Suosuo* shrubs establish by shooting out a network of roots that binds the earth to itself. Their dense foliage splits the winds at the turbulent surface, breaking nascent suspensions of sand and wind before they can coalesce into dust storms.

On his ex-pastureland, Mr. Li oversees this transformation of dune to shrubland, indeed even collecting forestry subsidies for his efforts. In his estimation, however, he is not a *suosuo* farmer. *Suosuo*, while a key species for anti-sand forestry programs, is not in itself economically valuable, he explains, even with the forestry subsidies that substantially underwrite its planting. For Mr. Li, it is another plant, another product, that drives this work. When *suosuo* establishes, its roots create the ideal micro-environment for *rou congrong*, a desert ginseng that, for its snakelike shape, is said to bolster virility through sympathetic magic.

While *rou congrong* roots thrive in sandy conditions, they only grow in the roots of plants like *suosuo*; that is, in order for *rou congrong* to root, it must inter-root. While Li has no interest in planting windbreaks per se, he nonetheless plants them anyway, as they are a botanical pre-requisite for the *rou congrong* that Mr. Li hopes will make his land prof-itable again. He tends these forests of shrubs as one builds and maintains a machine in a factory. '*Rou congrong*', he tells me, 'is a *chulu*', a road out of the dead end of his desertified pasture. And on this road out, growing and maintaining *suosuo* is the first step.

When Li looks out over his forest of seedlings listing in the wind, he sees money. Money put in, money borrowed, and accumulated through state incentives. He sees money in the future, when the shrubs establish and root production goes online, when he can sell these roots to state buyers, who promise stable prices and unquenchable demand – they will buy as much as he and the other ex-herders, on their patchwork of broken pastures, can produce. He sees money bet against his future, because even with careful maintenance, *suosuo* forests have a remark-able tendency to die before establishing and even after. The growing cycles of the plants comprise a multi-species business plan that links afforestation and root-harvest as moments in a broader economic calcu-lation that will make the sand productive again. *Suosuo*, he assures me, will turn the land into a biotic factory for these roots, spinning sand into gold.

As Li and other ex-herders retrofit the desertified pastures into future root factories, they have become inadvertent agents of the forestry bureau, growing the rows of *suosuo* that local forestry officials can and will count toward their afforestation quotas. The botanical pair of *suosuo* and *rou congrong* indeed makes the land, in the very same act of planting, *both* a forestry zone and a living machinery for root produc-tion, oriented toward a future market in roots. Alxa's Forestry secretary Ye, in his dusty office in Alxa's main town of 100,000, lays out this for-tuitous doubling of the forestland – in its infrastructural and economic functions – as the template of a political technique for absorbing herders into the implementation of state afforestation projects. 'If we can find ways to make herders convert their land to *rou congrong* farming', he says, 'they will grow *suosuo* more diligently than if we just paid them to grow windbreaks in the first place'. He continues, 'Those pastoralists will plant the *suosuo* forests and pay and work to keep them alive because they need *suosuo* to grow *rou congrong*.'

The inter-rooting of two plants in preparation for *rou congrong* cash cropping is an instance of what I call 'groundwork' in two senses, each fixated on the stabilization of the loose earth. For Li, planting of windbreaks is a groundwork in the sense of a preparatory labour that precedes what for Li is the 'real' work of root cultivation. Groundwork is what must be done so that things can get done. Second, for the region's forestry administration, it is 'groundwork' in the sense of a literal labour on the earth and its geophysical properties, a technical necessity for protecting downwind atmospheres from 'local' dust storms. Between these two kinds of groundwork, regional forestry officials see the promotion of monoculture root farming – or a di-culture of inter-rooted windbreaks and cash cropped roots – as a key means of achieving geophysical stability.

On degraded pastures, 'groundwork' stitches together changes in the region's economy as a possible means of generating a new sand-holding ecology. It is a way in investing in the unruly nature at desert's expanding edge as it shifts between two economic imaginaries and the social and ecological worlds gathered by action into earth on the move. Desertification and its control slips the broken pasture into an unruly motion, and in the continent-threading wind, state-sponsored overgrazing has made Alxa a meteorological frontier through land degradation. Where groundwork makes preparation and anticipation of new economic horizons into a nature made wild in transition, exherders and forestry officials inhabit the land as 'an edge of space and time: a zone of not yet', as Anna Tsing writes (2005: 28). This anticipatory affect, animated through the earth as a set of economic principles, geophysical transformations, and promises, is crucial as a governmental principle oriented toward future earthforms, markets, and opportunities, none of which have yet materialized (see Paprocki, this volume).

This engineered entanglement of economic and ecological changes, presaged fortuitously in the symbiotic inter-rooting of plants, makes economic and ecological processes disparate and yet interacting elements in a state-sponsored landscaping process. Groundwork, I want to emphasize, is a reminder that the hoped-for forestry ecology of roots, shrubs, and their planters, is not an end in itself, but rather a means of intervening into the geophysical landscape and its flighty properties. The multi-species ecology growing through the sand, that is, is the point where engineering an economy becomes a way of engineering a shifty landscape. It is the particular reworking of multi-species ecology into a potential technique of meteorological control that I want us to notice here. Life forms are openings to earthforms.

'Environment-making' names a governmental technique oriented toward instrumentalizing arrangements of things, acting on and modulating the conditions for new practices to emerge. Forestry officials maintain that it is important to build an effective economic *environment*, and not just a developing economy, to induce herders to plant. That is, human behaviours were thought as not only environmental, in the sense that they have effects that can be registered on some external world. This is the sense of culture that Marilyn Strathern describes as 'the workings of human activity as such' (2013: 215) whose effects might be registered in a scalar manner against some pre-given nature.

However, human behaviours are also 'environmental' in the sense that they could be elicited as reactions to changing, and therefore changeable environmental conditions. Behaviour in this sense does not emerge from an acting subject, but in reactions triggered by stimuli. Foucault's discussion of 'environmental technologies', associated with what he calls 'American neo-liberalism' (Foucault, 2008: 259) is a helpful way of framing this capacity of an economic environment to elicit behavioural responses. In his reading of a proposed measure to control drug use in the US War on Drugs based on liberal economics, the template of an environmental type of intervention is the artificial manipulation of drug prices as a way of manipulating the market milieu in which drug users operate. Prices were kept artificially high for new users who might yet be dissuaded by high initial cost, and they were tamped artificially low for habitual users so they will not commit criminal acts to secure cash for expensive drugs. Environmental intervention, he suggests, acts indirectly to efficiently condition behaviours. It acts 'on the market milieu in which the individual makes his supply of crime and encounters a positive or negative demand' (2008: 259).

In what follows, I explore how ecological construction programs more concretely experiment with market environments in order to dismantle pastoralism and institute a replacement economy for *rou congrong*. In this process, forestry officials and local cadres have experimented with building a carefully calibrated economic environment, from start to finish, aimed at creating the economic, labour, and market conditions for *rou congrong* and *suosuo* to propagate. This is predicated on an increasingly powerful governmental understanding that the tight management of specific economic conditions can orchestrate human economic behaviours in the creation and maintenance of state-sanctioned landscapes.

Bankrolling Symbiosis

In the wake of major storms in the early 2000s, a new meteorological villain burst onto the scene in Beijing. In accounting for dust storms, state officials blamed 'irrational land use practices' – most notably 'overgrazing' in upwind regions. In public discourse, the problem of overgrazing crystallized in the figure of ravenous upwind goats in distant dust source areas, transforming grassy pastures into deserts. Newspaper editorials in national publications demanded 'Killing goats and protecting grass' in Alxa and other points along the wind to protect Beijing's airspace. Cashmere goats, the mainstay of Reform Era pastoral economies in Alxa, emerged at the centre of a new form of atmospheric accounting, measured in dwindling grasses and increasing dust events.

Forestry officials and environmental NGOs in Alxa alike cite the conditions of an under-regulated market for animal products as the main driver of this degradation. They argue that the unregulated profitability of cashmere – 'soft gold' in local parlance – combined with the slicing of communal mega-pastures into smaller family plots meant increasing ecological and economic pressure on exhausted pastures. Herd sizes ballooned 10-fold in 20 years, in local accounts, in response to the high price of deregulated cashmere. The pastoral economy in its extant form for officials increasingly indicated a maladjustment of economic and ecological conditions and effects.

To protect grass and Beijing, mandatory seasonal controls on grazing were implemented in Alxa in the early 2000s. Several years later, the forestry administration offered full bans on grazing as a voluntary measure in the mid-aughts, funded by the nation's forestry agencies as anti-desertification work became one of their key mandates in northern China. Families afflicted by desertification – and whose herds were blamed for causing it – could, in exchange for their compliance with grazing bans, gain access to a suite of benefits and subsidies disbursed by regional, provincial, and national governments. Most immediately among the benefits attached to voluntary cessation of grazing were access to free and subsidized off-pasture housing, technical trainings, and access to streams of money for many new 'ecological' endeavours, including *rou congrong* plantings. It is these moneys that forestry programs would increasingly offer: not as emergency financial support in the aftermath of pastoralism but rather as a pool of starting capital available to ex-pastoralists to immediately reinvest in new, state-regulated economies retooled specifically toward state ecological construction goals.

For three decades after the dismantling of communal pastoral brigades in the early 1980s, the Li family raised ever-growing flocks of cashmere goats on the plot of land redistributed to their family after land reform, rapidly growing their herd to deliver the goats' meat and cashmere wool to a rapacious market for luxury goods in China and beyond. In 2010, Li's family, facing the dual pressures of desertification and the controls on grazing implemented to control it, decided to comply with these full state bans on grazing, selling 90% of the family's flock of goats in exchange for access to this package of government programs and payouts.

For their part, forestry mega-projects have cobbled existing institutions of socialist support and redistribution into streams of cashflow, ready for disbursement to prospective root growers. Additionally, state coffers are flush with new money, not only from well-funded forestry programs supported by the central government, but also money from Inner Mongolia's booming coal extraction industries (see Woodworth, this volume) – which are never blamed for the ecological devastation that has swept across pastures. When I speak with Li, he lists the government's programs each as a source of financing, culled into a fund that he uses as seed money for his family's new enterprises. But he also collected more money from other diverse state and non-state sources: state supplementary benefits for 'poverty relief' (*pinkun fuwu*), the windfall from the one-time sale of the family flock, loans from relatives and Alxa's regional agricultural bank, not to mention the combined pensions of he and his wife's retired, still-living parents.

'This year', he says, 'the first *suosuo* planting from two years ago will have grown large and deep enough, and we will begin the planting of *rou congrong*'. He interlaces his fingers to demonstrate the interlacing of the root structures of two plants, the one growing through the other. Li narrates his choice to participate in state forestry programs as essentially a matter of gaining access to new sources of investment capital, cobbling together state resources with other moneys. This money has underwritten the preparation of the land for *rou congrong* planting while getting the family through the two years before *rou congrong* can be interplanted.

Grazing bans and this reformatting of forestry into venture capitalism were part of a cluster of governmental techniques that aimed to induce pastoralist families to opt out of grazing, as if by choice. In order to avoid 'social chaos' in the aftermath of compliance of bans, forestry officials worked to create an immediate replacement economy that would 'catch' ex-herders in a new state-built ecological apparatus. A properly adjusted market could be used to reinvest pastoralists in their land's potential new productivity, while at the same time making pastoralists-turned-cashcroppers-turned-foresters into de facto arms of the forestry administration in their planting and maintenance of windbreak forests. It also aimed to create populations whose future economic livelihoods are directly dependent on the ecological construction programs that aim to stabilize dunes by transforming them into *rou congrong* farms, and thus *suosuo* forests.

Closing the economic loop, local officials have actively worked to create potentially infinite demand for *rou congrong*. They have promoted *rou congrong* as a local specialty product, experimenting with new uses for the root. Buyers from nearby Ningxia Province have recently set up shop here, coaxed by deep tax breaks and promises of political favour from the local government, which has invested in creating the conditions for a new resource boom in medicinal roots and other so-called sand products (*sha chanye*). As potential resources, these roots are not transparent receptacles of value waiting to be extracted, then. Rather, cementing their status as resources is the continual and explicit goal of forestry campaigns that depend on the creation of value in roots as a key spoke in the stoking of the economic and ecological environments deemed necessary for slowing dust storms in their tracks. The state creation, manipulation, and maintenance of an economy, Party Secretary Bataar insists, aims to remove any 'natural' economic barrier to the expansion of *rou congrong* planting, and thus to the *suosuo* forestry that it generates as its precondition.

Forestry officials work with other government agencies and companies in order to create a plug and play economy with niches waiting to be occupied by ex-herders. They are building a supply chain with everything but suppliers. Becoming a producer of the root is to occupy a structural position in a commodity chain anchored by buyers waiting, state money in hand, for an as-yet unproduced product. State manipulation continuously props open this market in roots, organizing economic activity without the pitfalls of unpredictable market fluctuations: an economy with guaranteed buyers meant that the market phenomena of over-supply, fluctuations in supply and demand, and even economic competition can be neutralized. Where more production does not mean falling prices, there can never be enough producers. Against Alxa's degraded ecology, this is a political economy of infinity rather than scarcity.

Cutting to the chase, when Li explains his decision to grow *rou congrong*, he points to the existence of these companies as the evidence of a state-guaranteed market. These conditions make him think of his land as the site of a long-term business strategy, one he narrates as a feedback system of mutually-causing economic and ecological transformations. Proceeds from sales will roll back into his *suosuo* planting fund so he can expand production, setting in motion the ex-pasture as a self-sustaining machine of step-by-step ecological and economic change. Because this market exists by sheer political will, it can persist as long as the state supports anti-desertification work, that is as long as Alxa remains a dust storm source area for Beijing, its earth a latent cloud. Against the endurance of the sand, these groundworks must continue.

Coda: Response

The status of Alxa as a meteorological frontier and the attempt to generate a resource boom as a state strategy of anti-desertification engineering characterize this frontier in a number of unexpected ways. These introduce a set of tensions and double movements into the notion of a frontier itself. First, as a dust storm source region, Alxa's remoteness from the centres of state power, as a sparsely settled, distant 'small place' (Kincaid, 1988) in the political geography of the country, is continually rendered as a meteorological proximity. As a key station in the formation and motion of a storm, this 'remote place' (Ardener, 2012; Harms and Hussain, 2014) appears in state forestry programs as a key chokepoint in a northern China figured less as a topography of windprone sandy lands, storm-channelling mountain passes, and plains that sprawl open as the bed of dusty rivers.

But in addition to this, Alxa and its hoped-for transition from pastoral sandscape to an infrastructure of dust-catching windbreaks in a bonanza of state-sheltered speculation means that frontier regions do not simply emerge in the freeing of a rampant capitalist impulse. Indeed, they emerge in the state promotion of market agency as a theory of human behaviour. In this last section, I attend to the meteorological frontier as an incubator for experimental techniques of social governance. I suggest that in the remaking of remoteness into meteorological proximity, the market, as an environmental medium for human behaviour, converts naturalized understandings of economic freedom and choice into new modes of social and ecological control, relation, and management. In doing so, the frontier has become a generative site for experimenting with the constitution of environmental subjectivity as an already constituted neoliberal subject. The *homo oeconomicus* presumed and continually enacted in environmental techniques of governing, however, does not appear either as the environment-loving affective and ethical subject presumed by Agrawal's formation of environmentality, nor as a rational agent seeking a competitive and individualized mode of economic advantage. Rather, this environmental economic subject is constituted as an ensemble of virtual behaviours that can be properly elicited by well-calibrated environmental triggers, as we will see below.

The creation of a highly artificial economic environment, bankrolled through the cobbling together of state moneys and driven by a demand held open by state intervention was posed as a means of orchestrating ex-herder behaviours. These interventions aimed to leverage the capacities of markets and other economic phenomena as tools for directing economic behaviours that would double as vehicles of desired ecological construction. At the centre of this is an idiosyncratic vision of the political subject that environmental techniques presume. This is an environmental subject that is quite different from the political subject of certain brands of environmentalism, for whom coming to 'environmental awareness' unlocks ecological change.

Curiously, any sense that there was a need to foster a care or consciousness of one's actions vis-à-vis 'the environment' was absent or incidental in any of these re-engineerings of the economic environment in service of ecological construction. In the case of anti-sand programs in Alxa, the environmental subject presumed by these techniques is one whose behaviours could be anticipated and elicited as *responses* to modifications to a milieu. It is necessary to foreground 'behaviours' and 'responses' as the politically expedient features of this figuration of environmental subjectivity. Foucault argues that in a becoming-environmental of power (Massumi, 2009), that the individual governable through action on its environment 'is sensitive to modifications in the variables of the environment and which *responds to* them in a non-random way, in a systematic way' (Foucault, 2008: 269, italics added).

The governability of this environmental subject inheres in its responsiveness to changed in its environmental conditions. That is, from the perspective of an environmental technology, the actions of environmental subjects can be interpreted as responses to environmental conditions. In Alxa, *behaviour* was the important level of response because behaviours could be coordinated for their effects on a physical environment. For instance, the means of converting ex-herders into de facto forestry workers through the dual planting of *suosuo* and *rou congrong* did not have to do with instilling or fostering a desire to do good ecological work or even a desire to plant. Rather, the creation of an economic environment that incentivized *rou congrong* production was aimed at eliciting a specific behavioural reaction – planting – without reference to subjectivity.

Curiously, the ecological environment is both central and strangely absent in Li's plans. While he describes his land in the midst of an ecological transformation, this is continually reframed as practice of economic preparation, a technical prerequisite for participation in the market in medicinal roots. Animated by a market in which demand for roots was propped open by political intervention, planting sand-binding plants takes on a distinctly economic character, better grasped through budgetary than biological considerations. Instead, his story of ecological transformation is subsumed in a story of navigating through, being a player in, a new economic game, where subsidies become seed money and shrubs become a means of production.

In any case the environment has 'worked'. In Mr. Li's plans, the ex-pasture is slated to become, with each season, an ever-expanding forestry zone. As he busily participates in the economic environment engineered to make ex-herders grow medicinal roots, he has coordinated the planting of shrub forest against dust storms. Variables in this 'environment' are available to perpetual strategic adjustment to elicit and activate the propensity of elements and in the fact of arrangement itself (Jullien, 1995). Effective configuration of elements in the economic environment in which human behaviours are epistemologically re-situated is a way of unlocking proper, controlled behavioural responses, which in this case can be adjusted to specific ecological ends. Mr. Li's participation in the *rou congrong* economy has been engineered into the economic environment as the anticipated response to well-calibrated arrangements of economic things and forces.

A crucial difference in economy and environment then, is this: where the economic agent acts, the environmental subject is configured as only ever having the capacity to *respond*. Or rather, in the framework of a technique directed at the control of behaviours through an economic environment, the environmental subject is one whose actions are always interpreted as a response to environmental conditions, and the environment is engineered in such a way as to activate economic agency as a response. When an environmental subject, so posed, acts, these acts are not a self-originating assertion of agency, but rather a confirmation of the efficacy of the conditions in which it is embedded. Crucial to the many explanations of ecological degradation in Alxa, and then to the programs to retrofit them, is that the environmental subject at stake in these techniques is one who never acts per se but who is seen as only ever responding, quasi-mechanically, to the changing conditions of its environment, in the format of a response to an introduced stimulus.

Notes

1 See *Renmin Bao*. (2000). *Zhu Rongji xialing 'shu yang hu cao' bao Beijing.* http://www.renminbao.com/rmb/articles/2000/8/23/2504.html.

2 See Pan, Yue. (2007). Shehuizhuyi Shentai Wenming (Socialist Ecological Civilization). Ministry of Environmental Protection of the People's Republic of China. www.zhb.gov.cn/hjyw/200702/t20070206_100622.htm.

Part II
Frontier Cultivations and Materialities

Part II

Framing Essay

Frontier Cultivations and Materialities

Nancy Lee Peluso

Resource frontiers are sites made up of and representing material and ideational transformations. They also constitute historical moments in which frontier-making processes, relations, and actors come together. In the course of assembling, they change material geographies, natures, and relations of power, control, and authority at work in these frontiers. Frontiers are sites in which the struggles over these changes are made visible in the formation, production, or extraction of these resources and the residual effects of these processes in and on the spaces of the lands or seas in which they are found. By the very 'nature' of their physical and ideational locations in historical trajectories of place-making, frontier assemblages embody difference – from other moments of assemblage and from other conceptualizations of frontiers.

It is not only the materialities of the resources or natures that are of concern in these three frontier studies; rather state actors' exercise of authority is key to understanding how and under what conditions state actors' actual controls on those resources and natures are established, realized, or abandoned. Though not all the frontiers described by the three authors in this section lie precisely at the territorial 'edges [of their] respective states' (Eilenberg, 2012) all three examine, to some degree, specific frontiers as edges of state authority and how that has been made, reinvented, or receded on the land or in the territorial waters of those states. While only Igor Rubinov focuses explicitly on state presence in the course of a remaking of frontier authority, Zachary Anderson and Heather Swanson also demonstrate how states can be formed or re-formed

Frontier Assemblages: The Emergent Politics of Resource Frontiers in Asia,
First Edition. Edited by Jason Cons and Michael Eilenberg.
© 2019 John Wiley & Sons Ltd. Published 2019 by John Wiley & Sons Ltd.

in what Keith Barney (2009) calls the 'relational space' of the frontier. Similarly, all three authors have something to say about the materialities at play as they are being or have been assembled at present and, in these chapters, at prior historical moments. The differences in the materialities of the resources at different moments, the shifts in their meanings, and the ways in which these are changed in relation to state interventions in extractive or productive regimes tells us something about the work that a 'frontier' label does under varied conditions.

If frontiers are relational categories and spaces of place-making, sites of struggle over resource control and representation, and moments in the formation of various sorts of state authority, frontier assemblages are emergent from the processes working with and against the material possibilities and constraints produced within those places. In Berau, Indonesia, Hokkaido, Japan, and the border area of Tajikistan, the authors address past and present moments of colonization and resource grabbing that have benefited national, colonial, and imperial state forms and institutions in ways that derive very much from the nature of the resources being overtaken. If frontiers are productive of wildness (Tsing, 2003, 2005), they are also sites where wildness is struggled over and tamed (Yeh, 2013); formal and informal institutions of control are imposed or emergent (Peluso, 2018; Rasmussen and Lund, 2018); and the state demonstrates its authority through labels of 'legal' and 'illegal' (McElwee, 2004; Erman, 2008), although the effectiveness of both institutions and proclaimed authority is not always realized. Emily Yeh (2013) in particular goes beyond the 'Wild West'-inspired depictions of frontiers to demonstrate how frontier subjects are formed, labelled, and disciplined in China's 'Great West' – and through the state's disciplining of subjects under different material regimes of practice and control, state territories have been realized in a space of fierce resistance to Chinese authority.

In Chapter 4, Zachary Anderson develops the process of 'frontierization' discussed by the volume's editors in one of Borneo's last districts containing significant amounts of mature forest. Drawing on imaginaries of Borneo that have long associated the island with vast rainforests, and extending this inherently 'green' vision to a subsequent era in which tree plantations are glossed as 'green development', Anderson focuses on two historical assemblages. These are differentiated by the material conditions of mature mixed rainforest versus tree plantations on what is or was state forest land – the political forest claimed by the Indonesian state. The landscapes of extraction/production constitute strikingly different frontiers and frontier assemblages in the same space.

In the chronologically 'first' frontier – and in the case of these forest spaces, chronology and forest 'age' partially constitute the frontier's material form and assemblage-formation – the imaginary of the pseudo-scientific 'primary' forest is one of the material resources around which an extractive assemblage forms. Notions of 'virgin' or inefficiently used land and seas inform justifications for 'improvement' (Williams, 1983) through development of oil, gas, and coal in both super- and sub-terranean realms (below soil as well as sea). The extractive regime both emerges from and leads to the state's declaration of its authority over all its national territory's resources (Moniaga, 1993); the creation of the political forest and the nationalization of fossil fuel resources; and, in the midst of all this development, the declaration of the spaces of Berau as a naturalized resource frontier. This is a frontier of the 1980s and 1990s – where logging and mining are tearing up the land and the coastal areas. This is the classic Tsing frontier: characterized by the extractive-ness of everything assembled around these particular resources (Tsing, 2003, 2005). State-sanctioned logging and mining proceed without regard to or recognition of the indigenous peoples and their prior claims and uses – making this extractive frontier into an Indonesian, Suharto-era correlate of Kerwin Klein's American 'Frontiers of historical imagi-nation' (Klein, 1999). In Tsing's Bornean frontier, these forested spaces are rendered wild, with the state and its cronies grabbing and extracting and ignoring previous occupants, destroying the forest civilization that preceded it as it creates the space as an extractive frontier. Though the processes of destruction in this frontier are highly dynamic, the longer term effects of frontier processes are less apparent. Focusing on that first frontier moment – the moment of wild extraction – unintentionally ren-ders this frontier as static.

In Anderson's piece, however, we are dealing with quite a different frontier moment. Here, states are still engaged but in other kinds of extraction that are glossed (and masked) as production. In this out-wardly tamed and systematized frontier, projects such as oil palm pro-duction, REDD+, the construction of smelters (for hard rock minerals), cleaning facilities for coal, and processing of oil and gas into commercial products belie an internal wildness. Here, the rights and claims of indig-enous and migrant producers remain unattended except through the lip service of conservation and industry groups touting their sustainability, their green-ness. This moment – the present – is a differently assembled conjuncture: different because of its materialities and different because of its vision of productive development and 'improvement' (Williams, 1983; Li, 2007b). From the point of view of those ignored early residents and new settlers alike, the space is still a wild frontier that is character-ized by impoverishment and dispossession of marginalized subjects.

Taming, disciplining, and a kind of re-wilding through the production of ruins (Stoler, 2013) are the themes underlying Igor Rubinov's intervention about the frontier constituted by the former Soviet Socialist Republic of Tajikistan. A frontier made almost entirely of mountains, and a symbolic bulwark at the edge of the Soviet 'empire' and on the brink of the [global] 'West', Tajikistan was once the symbol of Soviet greatness and power – the power to develop one of the world's most challenging frontiers. As also a tangible, visible, material gateway to the rest of the Soviet empire, Rubinov argues that Tajikistan was a frontier that garnered much more of the state's attention than other later-national frontiers along this edge of Soviet authority. It was literally through the construction of Soviet-style cement block edifices and the making of the high mountain landscape into a massive tree plantation, that the Soviets established and announced their 'superior' capacities to rule even the edges of a vast territory.

This assemblage of frontier power and production was not unique on a global scale, however. The discourses of scientific forestry were deployed here as they were extensively throughout the colonial world. The utterances of the Soviet foresters, and their creation of binaries to differentiate their improvements from the agriculture of 'destructive' and 'backward' locals, could have come from British colonial foresters (Bryant, 1997; Sivaramakrishnan, 1999); Dutch colonial foresters (Potter, 1988; Peluso, 1992) or their contemporary equivalents. What was different in these state forestry projects from those of the Soviet foresters was precisely the materialities of the trees planted: scientific foresters the world over have always been driven by the goal to produce *timber* from specific species of trees.

The mystery of the frontier fruit trees aside, in the present assemblage, Rubinov argues that local people in Tajikistan have been left largely to their own devices in the ruins of the abandoned model frontier. They improved, cared for, and expanded the production of the left-behind trees for their fruits rather than for their wood and erosion-containing purposes. The assemblage around this montane environment was now more constrained – and seems more localized in its material constitution. When the Soviets abandoned their frontier projects, the farmers and other mountain dwellers had to remake the infrastructures of Soviet authority to connect to an increasingly market-dominated region and world. While the move to producing high volumes of apples and walnuts was partly a result of the receding of the scientific foresters and the necessary initiative of the farmers individually or in associations, it was also a result of the material resources left behind in the mountains – thousands of fruit and nut trees – after the civil war.

In Chapter 6, Heather Swanson explicitly engages with both the material resources of a complexly-situated, and slowly transforming frontier in Japan. Focused on fish, but also on the contexts within which these fish thrive (the seas and rivers) and are processed (on land or on factory ships), Swanson builds on a political ecological tradition of explaining the difference that nature – or naturecultures and socionatures – makes to analyses of resource use and distribution (see, e.g. Suryanata, 1994; Swyngedouw, 1999; Zimmerer and Bassett, 2003; Goldman, Nadasdy, and Turner, 2011). The materialities and ecologies of the fish Swanson describes – wild and industrially farmed salmon – are at the heart of her story from the perspective of Hokkaido-Island-as-frontier-region-of-Japan. The Pacific salmon of the late nineteenth century, not long after the integration of Hokkaido into Japan, was to the frontier assemblage of that time and place what gold was to the California territory of the same moment. The valuable material resource led to a rush of settlers into the territory, to engagement with a still-in-formation nation-state, and to the eventual patterns of domestication of lands and or seas by changing the who, where, and how of producing fish or gold. The depletions that followed, the busts after booms, and the movement of resource production assemblages in these two frontiers were also somewhat parallel. But the outcomes for the respective resources' availability and productivity – and the roles of resources in question – were quite different. Whereas gold was either depleted or regulated to such an extent that extraction was economically unfeasible through industrial means, the salmon in Hokkaido actually increased. This unexpected outcome for these frontier fish derived from changes in the fish themselves, from the environments in which they lived, and from the assemblages through which they were associated with states, fishers, settlers, and all their varied activities. The entire frontier assemblage changed through the actions and activities of late-nineteenth and twentieth-century assemblage components, including the geographies of fish distribution. The marshlands and rivers that had been the fishes' homes, sources of food, and spawning grounds were transformed by the development and drainage of these waterways and liminal water-land marsh environments. Northern and Southern fisheries on the island; wild catch versus hatchery production; and the biological characteristics, habitat locations, and ecological contexts and practices of the fish are diversifying, changing, and diverging. Though fish quantities have increased overall, production quantities, costs, and value differ – as does the resulting capital available in different localities for reinvestment. The resulting assemblages around fish production in different parts of Hokkaido, are thus themselves differentiated and more precarious for some fisher communities.

The frontier materialities around the various cultivations described by the three authors writing in this section differ from one another not only because of their 'resource' socionatural or naturecultural forms as forests, fossil fuels, fruit trees, mountains, seas, and fish, but because of how they are positioned within historical trajectories of frontier assemblages. Frontiers will always differ materially both due to physical geographies, ecologies, and locations, and to the historical moments or conjunctures within which they come into being as assemblages. In all three of these chapters, naturecultures or socionatures influenced what was possible under particular conditions in specific frontier spaces – or by helping constitute those conditions. While it may still be difficult to document non-human intent in these circumstances, the assemblage as a multi-component process enables the recognition that human actions and institutions do not produce resource frontiers on their own.

4

Mainstreaming Green

Translating the Green Economy in an Indonesian Frontier

Zachary R. Anderson

Introduction

In November 2009, at a meeting of the Governors' Climate Change and Forests Task Force in Los Angeles, California, the Governor of East Kalimantan, Awang Ishak Faroek, publicly pledged to make his province 'green', in alignment with greenhouse gas (GHG) emissions reduction commitments made by the Indonesian President in the lead-up to the 13th UNFCCC COP. This pledge resulted in the establishment of the *Kaltim Hijau* (Green East Kalimantan) program in early 2010 as an effort to curb the social and ecological threats posed by climate change and implement 'green development' in order to realize a 'just and prosperous society' (Province of East Kalimantan, 2011).

Governor Awang's vision calls for a radical transformation of East Kalimantan's economic base, from the existing 'black economy', based on non-renewable resource extraction, towards a 'green economy' based on supposedly more sustainable agricultural production, manufacturing, and the provision of ecosystem services, especially carbon, through the development of a provincial REDD+ pilot program.[1] However, somewhat paradoxically, this new green economy, is predicated on a 1.2-million-hectare expansion of the province's oil palm plantation area and investments in biofuel processing, both of which may lead to further deforestation (Anderson et al., 2016), and ecological degradation (cf. Cramb and McCarthy, 2016). Oil palm is the most productive oil seed crop in the world, and palm oil is used in a range of

Frontier Assemblages: The Emergent Politics of Resource Frontiers in Asia, First Edition. Edited by Jason Cons and Michael Eilenberg. © 2019 John Wiley & Sons Ltd. Published 2019 by John Wiley & Sons Ltd.

food, cosmetic and industrial applications. More recently palm oil-based biodiesel has been marketed as a sustainable biofuel to supplement and (eventually) replace fossil fuels. Yet, the spread of oil palm plantations in Indonesia and elsewhere across Southeast Asia, has been tied to extensive deforestation, negative impacts on biodiversity, and the displacement of communities from purportedly 'degraded' forest areas (Gibbs et al., 2010; Koh et al., 2011; Wicke et al., 2011; Gunarso et al., 2013; Margono et al., 2014).

This shift in East Kalimantan's development discourse is indicative of the region's changing position within Indonesia's national energy-resource nexus and exemplifies processes taking place in resource frontiers across Indonesia and Southeast Asia more broadly (cf. McCarthy et al., 2012; Fox et al., 2014; Eilenberg, 2015). East Kalimantan is one of the most sparsely populated provinces in Indonesia, but also one of the richest, with a natural resource-based economy that has at different moments supplied 50% of the coal used domestically (JATAM, 2010)[2], and up to two thirds of the country's timber production (Magenda, 1991). However, despite this abundance of natural resources, East Kalimantan continues to be characterized by marginalized communities and endemic poverty. As the province's oil, gas, and coal reserves dwindle and national-level discourse shifts towards low-carbon energy production and emissions reductions, the provincial government has embraced the concepts of 'green growth' and 'green economy' as strategies to bring 'development' to local communities while maintaining the province's important position in the national economy. The emergence of the 'green' economy in East Kalimantan is, at least partially, a manifestation of the government's desire to shift the province's role, nationally and internationally, from being a resource frontier dependent on coal, oil and gas, and logging, to a frontier for innovations in 'green' resource provisioning and development practice (see also Günel; Paprocki; Zee, this volume).

The district of Berau, located in northeastern East Kalimantan, has become a testing ground for this shift, with numerous actors and organizations working to realize their vision of green growth at the jurisdictional scale. The district government has capitalized on growing international interest in tropical forests as sources of ecosystem services to obtain support for capacity building and improved territorial governance, while transitioning the district economy from coal to an agro-industrial economy based on oil palm. Contested, 'development', in the guise of green growth, has become the hegemonic discourse around which various state and non-state actors plan Berau's future (see also Zee, this volume).

The Green Economy and Frontiers of REDD+

This chapter explores the ways in which frontier imaginaries and new modes of green capitalism are entwined and negotiated in the production of Berau's 'green economy' as a particular moment of frontier assemblage. While there is no universally agreed upon definition, 'green economy' is generally presented by proponents as an apolitical and 'eco-rational'[3] blueprint for 'sustainable development' – encompassing a 'triple-win' for environmental sustainability, poverty alleviation and economic growth (OECD, 2011; UNEP, 2011). At its most basic, the green economy is seen as a process for harmonizing economic growth with environmental sustainability, and a way to 'decouple' economic growth from the environmental and social ills that have characterized economic development over the last century (see also Choi, this volume).

As the debate around REDD+ and the international carbon market has dragged on, implementing organizations, local governments and communities have become increasingly sceptical about the likelihood of carbon payments reaching project sites. In Indonesia, this has contributed to the embrace of the concept of 'green' economy as a means of pursuing emissions reductions across sectors, rather than focusing specifically on reducing deforestation and forest and peatland degradation. This shift from REDD+ to 'green' can be seen in the development of the Berau Forest Carbon Program (BFCP) – a district-government-led REDD+ initiative designed and managed by The Nature Conservancy (TNC) and implemented in collaboration with the German Development Agency's (GIZ) Forests and Climate Change program (FORCLIME). Although it was started by TNC and is technically a REDD+ demonstration activity, in practice the BFCP acts as an umbrella for all 'green development' activities taking place in Berau, with the Berau REDD+ Working Group serving as manager, and TNC as the lead implementing partner.

What I call Berau's green economy is a historical, spatial and analytical assemblage that transcends easy demarcation by scale, time period or a stable network of actors (see also Middleton; Swanson, this volume). Berau is, of course, a material place with its own unique landscapes, eco-systems, history, and cultures; however, 'Berau' also exists more abstractly within the imaginations, policy documents, and project descriptions of various actors and organizations. The concept of 'frontier' as developed by Tsing (2003, 2005) and others (see Cons and Eilenberg, this volume) is useful in exploring how the 'green' economy is being translated across scale and between actors within this imaginative landscape, and how dominant knowledge and discourses about Berau' forests, populations, and resources are produced and sustained through a particular process of 'frontierization' (Cons and Eilenberg, this volume) tied to a form of 'green' neoliberal capitalism which has become hegemonic in conservation and development circles globally (see also Paprocki; Günel, this volume).

Berau has undergone two intertwined but distinct moments of fronterization, which led to two moments of frontier assemblage. Historically, Berau has been assembled as frontier for resource extraction – a forested space full of undeveloped resources, imagined to be empty and ready to receive migrants from across Indonesia. However, as capitalism has shifted to incorporate the concerns of environmentalism and neoliberal governance, the frontier assemblage in Berau too has shifted, and what we are currently witnessing is a moment of fronterization in which Berau is positioned as a frontier for sustainable development and the 'green' economy, yet in which production remains extractive in nature. While continuity exists between these two moments of frontierization, particularly in the forms of production and resource extraction driving economic growth, what sets Berau's contemporary 'green' frontier assemblage apart is the role of non-state actors in governance activities, and the ways in which efforts to govern Berau's peoples and landscapes have increasingly relied upon biopolitical and neoliberal governmental practices and technologies.

As Tsing (2003, 2005) suggests, frontier expansion is not a matter of taking a natural entity and fitting it into existing networks of global capital but rather represents a unique and complex form of capitalism in itself. Thus, this contemporary moment of frontierization is itself integral to shaping the form that green capitalism and 'green economy' will take across Indonesia, as this model of development is 'scaled-up' and transferred to other areas across the country. I argue that Berau's green economy can be seen as a unique form of frontier capitalism in which the oppositional discourses of environmentalism and indigenous rights have been enclosed and brought into the service of capital under the guise of sustainable development.

However, while the concept of the 'green' economy has gained stature within Indonesia's national development discourse and amongst international development actors, the policies associated with the green economy are not simply formulated by a specialist community of state and non-state actors and then filtered down, unaltered, to the local level. Berau's green economy is being constituted through the convergence of a diverse network of actors and institutions around efforts to implement and 'demonstrate' 'green' reforms of environmental and territorial governance. The framing of conservation and governance reforms in terms of free market eco-rationality has been key to acquiring international funding and support for the projects of the green economy in Berau, yet the diverse concerns of local government and community actors have also been important in determining the form that Berau's green economy has taken.

Because the green economy, and associated market-based conservation projects, as envisioned by international donor agencies and environmental NGOs, often diverges from the vision of development held by local governments and forest-dependent communities, implementation of these projects and policies may lead to disappointments and unmet expectations (see McGregor et al., 2015). Thus, as the projects and policies of the green economy travel across scale, moments of unplanned-for 'friction' emerge (Tsing, 2005), particularly as efforts are made to insert and define local spaces into new and existing networks of global capital, as frontiers of extraction and experimentation (Tsing, 2003). These frictions are integral to shaping the form that the green economy has taken in Berau, as various actors embedded within Berau's existing political economic structures negotiate and translate the various visions of 'green development' espoused by national and international actors. The result has been a green economy characterized by a vision of modernity and development in which resource-extraction-based profit accumulation remains ascendant, and land-use planning reforms operate to make previously 'unproductive' and inscrutable spaces and populations legible to the state and global capital (see also Zee; Choi, this volume).

In her work on community forestry, Tania Li (2007a) highlights a set of 'practices of assemblage' that are useful in interrogating the social relations underlying the contemporary period of frontierization and the emergence of the green economy in Berau. In this analysis, I build on and modify these practices using the concept of translation, as discussed by Michel Callon (1986) and Bruno Latour (1986). In this framework, translation is understood as a process through which a chain of enrolled actors can shape and change a token – such as a policy, concept, or claim – in accordance with their own projects and interests, rather than simply resisting it or transmitting it in its entirety (Latour, 1986). This can involve, but is not limited to, linguistic translation and the translation of meaning, such as the meaning of 'master metaphors' (Mosse, 2004), as well as the negotiation and demarcation of the identities of actors and the arena of interaction (Callon, 1986:203). What the green economy comes to mean in Berau will necessarily be the product of interactions between the new forms of knowledge, value and governance associated with the international project of green economy and the existing, place-specific identities, alliances, and social structures that exist in Berau. The practices of assemblage discussed in this chapter are:

problematization and authorizing knowledge, enrolment and forging alignments, and *rendering technical and apolitical.*

Assembling a Green Frontier – Problematization and Authorized Knowledges

Berau's green economy began to form as a coherent project of jurisdictional governance in the wake of the 2008 UNFCCC COP13 and the rise of international and national discourse about REDD+, and thus is intimately tied to the idea of using carbon forestry and improved forest management as means to reduce GHG emissions. The central proposition of the green economy, as discussed in meetings and interviews in Berau, is that economic growth, while imperative, must be decoupled from environmental destruction, particularly deforestation, water pollution, and GHG emissions. Proponents claim that this can be achieved through improved technology, better business and resource extraction practices, and improved spatial planning by local managers at the district and village level, particularly in relation to forests. This discourse identifies a lack of data as a key problem underlying environmental problems in Berau, rendering climate change and state-failure as technical matters. Data standardization projects and collaborative 'visioning' activities are regularly presented as the best solutions to tackle the environmental 'externalities' of economic growth, signalling an embedded faith in the 'rational' implementation of technologies to address the disconnect between economic growth and ecological protection. As one high ranking official involved in national development planning said, 'we cannot even begin to think about dealing with problems related to corruption or enforcement before we have accurate data to work with … there will always be corruption here and there, but with clear data local governments will be able to plan better and can be held accountable by the provincial and national governments'.

   TNC's non-confrontational, pragmatic, and market-based approach (TNC, 2015) has been instrumental in shaping the BFCP, and Berau's green economy more broadly. At the same time, because the BFCP was intended to be a project led by the district government, the priorities, concerns, and sensitivities of the local government have translated the concepts and intentions of the BFCP in a number of ways that, at times, appear adversely related to its original stated goals. For example, while the BFCP is at its core a REDD+ project that will ostensibly be supported by the sale of carbon credits, in Berau there is rarely any mention of the carbon market, and REDD+ is seen as a means of improving local capacity and forest governance rather than a means of creating saleable ecosystem service credits. This is at least partially due to an unwilling-ness of certain district government actors to engage with a carbon market program because of previous disappointments related to a proposed CDM afforestation/reforestation project that never came to light.

Officially TNC identifies the main drivers of deforestation in Berau as '[conversion of] land to oil palm and timber plantations, commercial logging, coal mining activities, [and] swidden-fallow agriculture' (Fishbein and Lee, 2015: 11), yet, until the advent of a recent sustainable oil palm project, the BFCP had virtually no engagement with oil palm or mining companies, which account for 44% of deforestation in the dis-trict (Casson et al., 2015). This problematization is further complicated by the fact that GIZ FORCLIME has identified the main driver of defor-estation in Berau between 1990 and 2010 as shifting cultivation and smallholder agriculture (Navratil et al., 2012; Bellot, 2015). However, as an informant working for a community development NGO in Berau pointed out, this can at least partially be explained by the fact that as oil palm plantations have expanded into village areas, communities have generally had to give up existing agricultural land, and thus must relo-cate their crops, causing a substantial increase in indirect deforestation. Even GIZ's study shows that after 2005 plantation development became the driving force for almost a third of all deforestation taking place in the district (Navratil et al., 2012).

In Indonesia, decisions driving land-use change and deforestation gen-erally happen far outside of the spaces in which deforestation and land-use change actually take place, with government decision-makers and land-use planers. Therefore, initiatives which focus on changing village-level land-use generally only target the symptoms of deforestation, rather than the underlying drivers (van Noordwijk et al., 2013). Nevertheless, while most of the donor and NGO projects supporting Berau's 'green' economy have worked to develop government capacity and integrate 'green' cost-benefit analysis as part of the planning process, they have also implicitly placed most of the blame for deforestation and forest deg-radation on forest-dependent communities, as the 'village', the 'farmer', and 'swidden agriculture' are constructed as the primary sites of inter-vention, against the backdrop of a supposedly untouched wilderness.

The experiences of a man named Siang[4] are illustrative of this issue. Siang lives in Gunung Madu, one of the BFCP 'model villages' identified in Figure 4.1 and has at various times been more or less involved in different 'green' projects, including those supported by the district government and TNC. As part of Gunung Madu's engagement with the BFCP, a conditional payment agreement was made that gives the village infrastructural and land-use planning support, as well as funding and resources for livelihoods programs and capacity building, in exchange

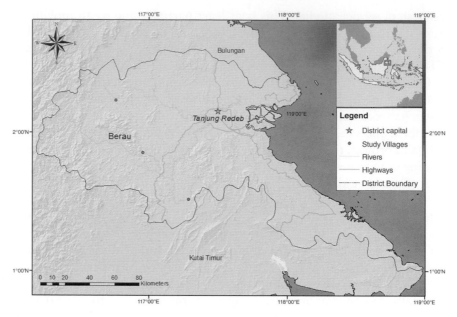

Figure 4.1 Overview of the district of Berau. *Source*: © Zachary R. Anderson.

for villagers limiting the amount of forest they can clear for subsistence agriculture. This includes limiting the maximum number of 0.5–1 ha *ladang* (dry rice cultivation) plots per household to four, and only allowing new plots to be opened in certain areas which are in agreement with the new village spatial plan. In the beginning, Siang saw this agreement as a welcome source of support and was not worried about the new restrictions as his own cultivation generally fell within the agreed upon limits. However, as oil palm plantations have expanded in the areas adjacent to the village's land, farmers from a neighbouring village have begun to encroach, opening *ladang* on land that Gunung Madu claims as its own, based on traditional land use and existing government spatial plans. As a result, farmers in Gunung Madu have opened *ladang* in this area with the goal of protecting their village's territorial integrity. Yet, because these plots were not specified in the village spatial plan there was concern that the village had not held up its end of the conditional agreement (Figure 4.2).

Following an evaluation of the village conditional agreement, Siang was clearly frustrated and identified what he sees as a disconnect between the goals of the BFCP and the focus of its activities. As he said, 'it is confusing why we have to justify our need to open new land to protect our village area and feed our families when the forest beside us is already

Figure 4.2 A truck hauling oil palm fruit bunches from a plantation near the village boundaries of Gunung Madu, one of the BFCP model villages. Photo: Zachary R. Anderson.

disappearing, being converted to oil palm. We want to protect the forest, but we have to eat so we open up a little land. But we are not the threat. The threat is coming from outside – from oil palm and logging. We appreciate TNC's help, but many of the [livelihoods] programs have failed – our chickens have died, as did the fish. We have attracted a lot of attention and many visitors are coming here to see our forest and mountains, but there are not many opportunities for a farmer like me'. Alternatively, an NGO representative familiar with Gunung Madu suggested that the opening of these new *ladang* plots has less to do with protecting the village than with individual farmers attempting to lay claim to the threatened area in anticipation of compensation if this area is converted to oil palm. Yet, the disconnect between the targets of the BFCP and the large-scale drivers of deforestation, as Siang identified them, remains.

Enrolment and Alignment: A Process of Translation

Development projects such as the green economy need to enrol a variety of supporting actors with reasons 'to participate in the established order as if its representations were reality' (Sayer, 1994: 374). There are a range of actors enrolled in Berau's green economy, each with their own priorities and objectives, and each instrumentally positioning themselves to capture new flows of funding associated with the construction of Berau as a frontier for the green economy. A key dimension of these actors' enrolment is the construction of 'mobilizing metaphors' (Mosse, 2004) that reconcile disparate perspectives around common goals and strategies. The green economy is a particularly potent catalyst for this alignment, in its presentation as a four-way win – being 'pro-poor, pro-job, pro-environment, and pro-growth' (GoI and GGGI, 2013)[5]. The all-encompassing nature of these claims allows the green economy to be presented as apolitical and technical – a project with something for everyone – concealing underlying ideological differences and reconciling diverse visions of 'success' in order to build a coalition of distinct interests (Mosse, 2004: 663). This process can be seen in the enrolment of the different actors discussed below.

The overall goal of the BFCP is to develop 'an integrated, jurisdictional-scale approach to green development that can serve as a model for how Indonesia can meet its ambitious economic development goals while reducing emissions and environmental impacts compared to business as usual' (NGO expert interview). This amorphous umbrella allows a range of activities and actors to be brought together under the vague goal of utilizing the market, governance reform, and public-private partnership to reduce emissions, make conservation pay for itself, and improve the well-being of local communities and the economic development of the district.

FORCLIME, TNC, and the other NGOs operating under the umbrella of the BFCP take 'community-based' participatory approaches. A major premise underlying their work is that communities' ability to manage forests is latent, and with proper support, communities will choose to manage forests in ways that reduce deforestation and protect biodiversity. Thus, the benefits promised to local communities are tied closely to attempts to govern their conduct. For communities that can be 'responsibilized' (Rose, 1999) to exhibit the right characteristics and carry out the right actions, the BFCP provides an opportunity to gain access to previously inaccessible government and donor development funding, NGO livelihoods subsidies, and support for community mapping. Moreover, alignment with the BFCP offers communities the chance to legalize their access to forest land through *hutan desa* (community forest) designation, and gain political influence locally and regionally. With TNC's help Gunung Madu was recently granted management authority over the 10,000 ha *hutan desa* within its boundaries.

A number of villagers from Gunung Madu, including the village leader and the head of the village forest protection organization, have become deeply enrolled in Berau's green economy, spending much of their time in the district capital Tanjung Redeb at various meetings and workshops, serving as the voice of the 'local people' in 'greening' activities. This role has extended internationally, as Gunung Madu serves as an Indonesian model of REDD+ 'success' and hosts visits from key international actors such as Norway's climate and environment minister and ambassador to Indonesia, as well as a range of journalists, ministers from Jakarta, and representatives of other donor agencies.

Frederik Ombu, Gunung Madu's leader, is both charismatic and adept at presenting an image of Gunung Madu that fits national and international discourse. In his own words, 'before we began working with TNC our forest was mismanaged, and people [from the local community] were opening land anywhere they wanted. The threat of outsiders logging, and oil palm entering our village was constant. But there have been many benefits from working with the BFCP. We know more about how to manage the forest and are in a stronger position when dealing with the companies [logging and oil palm]. We also have many friends in the Ministry of Forestry now, and internationally; many people are coming here to visit. The livelihoods programs are useful to get people interested in the program, and to help us be independent, but the most important benefit we've gotten is our stronger position when dealing with the government and companies.'

This process of enrolment is neither unproblematic nor complete, as pre-existing loyalties and objectives remerge, and divergent goals manifest themselves. As in other places, short project funding cycles can lead to discontinuity and unmet expectations, yet East Kalimantan is unique in that a network of non-governmental environmental practitioners and activists has been active in the province since the late 1980s, centred around the Mulawarman University Faculty of Forestry. This is not because this faculty has been particularly radical, but rather because, as has been discussed elsewhere in Indonesia (cf. Tsing, 2005), during the 1980s and 1990s in East Kalimantan organizing around environmental issues was often used as a covert means to address a range of social issues, including indigenous land rights, and resource access and distribution. Funding for environmental campaigns coming from major international donors like USAID helped to create a network of fledgling organizations, often created and managed by students and graduates of Mulawarman University Faculty of Forestry. Today the majority of the individuals working in international NGO and donor organizations at the district and provincial levels come from within this network and move between organizations as funding, projects and life dictate. Because of this, there are strong links between the different organizations and a shared understanding of key concepts and strategies has developed over the years, leading to continuity within this larger network that translates, at least partially, into Berau's green economy.

Additionally, the BFCP is an open-ended project, and while the activities grouped under it may change, so long as the government and TNC continue to support it, continuity remains. However, this support, particularly from government actors, is not guaranteed, as the BFCP is not sufficiently institutionalized within the local government and remains grounded in the fragile social relations of a network of enrolled actors. Moreover, the BFCP's durability to date is at least partially attributable to the program not having posed a meaningful threat to resource extraction in the district, nor to the existing networks of power, and the licit and illicit sources of revenue linked to resource extraction. Thus, paradoxically, if Berau's green economy becomes more successful at meeting the conservation goals described by some actors and limiting resource extraction-driven deforestation, it may risk losing the support that underwrites its existence.

Rendering Technical and Apolitical

Practices of rendering technical and apolitical highlight the work that must be done to bound Berau's green economy and construct it as a frontier, both geographically and discursively. The complex political, social, and economic 'messiness' of deforestation and environmental change must be simplified in order to make sense as a site of intervention, and both state and non-state actors must work to translate existing environmental problems and political economic conflicts into linear narratives resolvable by expert diagnoses and technical interventions (see also Paprocki, this volume). While this is true of development practice more broadly (Ferguson, 1994), the green economy represents a novel confluence of dominant development and conservation perspectives and draws its legitimacy from a supposedly economically and ecologically rational approach to development which represents the capture of both moral and economic challenges to development practice on environmental and social justice terms. In practice, this narrative is translated through hundreds of workshops, presentations, policy documents, capacity-building trainings, and monitoring trips, all designed to discursively produce Berau as a frontier in which green development interventions can, and are, remedying environmental governance problems and the unevenness of economic development in the region.

As with REDD+ and emissions reduction programs more generally, the deployment of technical and scientific concepts and vocabularies, most of which are poorly understood by people outside of conservation organizations, has become a powerful tool for legitimizing and envisioning the green economy (Astuti and McGregor, 2015:26). In the setting of the BFCP the terms associated with constructing the green economy as a bounded, technical domain are complex and defy easy definition, particularly as they are translated from English into Indonesian, or not. Terms like participatory land use planning, safeguards, conditional payments, carbon accounting, certification, and acronyms like Reference Emission Level (REL), Measuring, Reporting and Verifying (MRV), and Reducing Emissions from Deforestation and forest Degradation (REDD), are drawn from the English language, and rarely translated, in speech or in text, even when local equivalents exist. In this context, their unmoored nature allows them be attached to new meanings shaped by the interactions between different languages, people and forms of knowledge (Tsing, 1997). As Tsing (1997: 253) notes '[m]eaning arises from the slippages and supplements of the confrontation'. These terms, and the individuals, organizations and programs that use them, hold significant power in constructing the green economy as a technological solution to the problems of development, the environment, and poverty (Figure 4.3).

Figure 4.3 A 3D model of Song Kelok, one of the BFCP model villages, and the surrounding environment, developed through a participatory mapping process led by TNC. Markers identify existing land and resource rights, right holders, important landscape features and the village boundaries. Photo: Zachary R. Anderson.

The technical descriptions of the BFCP often present simplified narratives of problems/solutions that gloss over existing tensions, making the program appear far more coherent than it actually is. As Berau's 'green' economy has developed its problematization and field of intervention has become more complex and encompassing; from the management of communities' land-use practices to the negotiation of complex private-public partnerships and the modelling of the emissions associated with different land-use scenarios. However, one thing that all of the projects grouped within Berau's green economy share is an intention to 'verifiably' reduce GHG emissions, and thus require the establishment and recognition of a historical emissions baseline, or 'reference emission level' (REL), and/or a hypothetical 'forward-looking' projection of future emissions, against which interventions can be measured. The goal in Berau is to develop a district-wide 'carbon accounting framework' that takes into account emissions from all activities and land uses, thereby avoiding displacement of emissions-generating activities to other locations (at least within the district) (Fishbein, 2010). Although this may seem like a technical matter, it overshadows a range of complex social and political issues, homogenizing Berau's socionatural landscape under the lens of carbon accounting (Robertson, 2012). This process also further entrenches Berau's frontier status as contemporary extractive patterns tied to Berau's status as a resource frontier for boom crops are projected into the future as timeless 'business as usual'.

Much of the funding directed to 'green' initiatives in Berau has gone towards establishing this REL, and at present there are two versions; one created by TNC and one by GIZ. During interviews, informants argued that, despite coexisting, these two RELs are not actually in conflict, because they use different methodologies and highlight different aspects of emissions planning which can benefit the local government and project implementers in different ways. However, to date, the district-level government has been unwilling to select one official REL. There are a number of reasons for this, including a lack of clarity regarding the different methodologies, uncertainty about the likelihood of future emissions reduction payments, and fear that setting a REL will bind the district to reduced economic development. Yet what is clear is that the selection of a REL, which is presented as a straightforward technical exercise, is actually rife with politics and confused by different techniques and presentations tied to divergent ideologies of development and conservation.

Conclusion

At the village level, communities in Berau find themselves at the confluence of multiple forces, attempting to negotiate the promise of future benefits associated with REDD+ against the more tangible benefits associated with oil palm, while attempting to protect village forest areas and traditional ways of life. The same can be said of those in Berau's district government, who recognize that resource extraction as it has existed in the district for the last forty years is unsustainable yet find few durable alternatives to bring development to the district. As a project that rests on the logics of neoliberalism, the green economy must compete, economically, with other existing and new projects, many of which drive environmental destruction through support for agricultural expansion or other forms of resource extraction. Thus, unless the projects of the green economy are able to pay more than these industries, expectations of long-term durability and structural change are unlikely to be met. Yet, what I have hoped to illustrate is that this is not a simple technical matter to be resolved through 'better implementation' or by 'paying more', but rather a reflection of the core contradictions inherent in the model of the green economy and the simplifications associated with rendering Berau as a green frontier.

Berau's green economy is presented as an already existing model of 'green growth', and a frontier crying out for investment and development. Yet, this discussion of the situation on the ground illustrates the geographically and contextually-specific negotiations that characterize the green economy as it actually exists. In the green economy, we see the reproduction of the crises and contradictions that characterize capitalism more generally, and the continuation of issues that have plagued development projects in Indonesia for decades. The disconnect between district and village-level spatial planning, the ad hoc distribution of resource extraction concessions, and the superficial nature of current environmental assessment processes all have complex social and political dimensions which will remain unresolved if approached as matters only requiring 'better data' or more 'participatory' processes.

The argument presented here is thus not that the green economy is something entirely new – a drastic departure from conservation and development policy to date. In fact, it is the opposite. While being presented as a revolutionary, all-encompassing, and scientifically rational means for pursuing economic growth while addressing poverty reduction and ecological collapse, what we see in practice is the continuation of sustainable development, as it has operated since the late 1980s, having fully internalized the logics of neoliberal conservation. This observation, in itself, is not radical, and the ideology and discourse of the green economy has been critiqued on these terms (Brand, 2012; Bailey and Caprotti, 2014). Yet, to date, few examples have investigated the messiness and frictions inherent in the roll-out of the green economy in the frontier spaces which play a central role in debates about climate change and forest-based mitigation and adaptation. While the concepts of decoupling and offsetting are central tenets of the green economy, what is not often discussed by its proponents is the equity of 'fixing' climate change mitigation in the peripheries of the Global South. Rather than a revolutionary and forward-looking demonstration of social and economic change, what we see in Berau is the negotiation of limited options, the hopes of a range of actors for a better future, and the pragmatic acceptance of the best possible deal for right now.

Notes

1 REDD+ refers to a program designed to mitigate climate change through reducing emissions of greenhouse gases through enhanced forest management in developing countries, which is being implemented by the UN-REDD Programme and the Forest Carbon Partnership Facility (FCPF).

2 JATAM (2010). *Deadly coal: Coal extraction and Borneo dark generation*. Jakarta: Jaringan Advokasi Tambang (Jatam).

3 'Eco' here refers to economic and ecological rational actors (cf. Goldman, 2001).

4 With the exception of Berau, all of the names used for people and locations in the following sections are pseudonyms.

5 GoI and GGI (2013) *Government of Indonesia and GGI Green Growth Program: Prioritizing investments: delivering green growth*. Joint Secretariat GoI-GGI Green Growth Program. BAPPENAS. Jakarta.

5

Growing at the Margins
Enlivening a Neglected Post-Soviet Frontier

Igor Rubinov

Introduction

The sound of hammering pierced the quiet of a crisp fall day in the Pamir mountains of Tajikistan. I found Adham[1] working in a rough-hewn shed behind his home. As I stepped in to observe his handiwork, he admitted his craftsmanship was limited: 'Everything is a bit crooked, but I don't care. I just need to keep the mice away from the apples through the winter'. He was retrofitting his shed to store the longer-lasting, foreign apple varietals that he had started cultivating. As I helped shuttle in knotted planks of timber he had grown, felled, and planed himself, Adham lamented how rudimentary his efforts felt. 'We were once part of a great empire. We had in-ground heating centuries before it appeared in Europe. Then, in the twentieth century, we took part in a huge socialist project. So much changed so quickly. But what have we achieved as a country since those times? We are surviving on relics of the past'. Adham, a former manager of the collective farm (*sovkhoz*), had taken over its orchards and grafted new scions onto Soviet roots. When socialism faded away, the residents of the frontier enlivened the ruins of past projects to transform the landscape.

The Tajik Soviet Socialist Republic (SSR) was positioned on the edge of the Soviet empire, wedged against Afghanistan and China. The mountainous terrain of Tajikistan (cradling the highest massif in the Soviet territories, Peak Communism) comprised the furthest reaches of a Tsarist era buffer zone. The region served as a check on the British empire's

Frontier Assemblages: The Emergent Politics of Resource Frontiers in Asia,
First Edition. Edited by Jason Cons and Michael Eilenberg.

ambitions, setting the stage for a spy war later popularized as The Great Game. As Soviet rule solidified, these marginal lands became a showcase for its ideological commitments. Remoteness proved a virtue. The Tajik SSR emerged as an important platform to validate the global appeal of the Soviet Union's modernizing project. This immense effort took decades to carry out. Large-scale mines, tortuous highways, and towering factories were financed to usher the frontier toward industrialized modernity and tie the region to the Soviet core. However, the benefits of this largesse disappeared seemingly overnight. A horrific civil war came on the heels of Tajikistan's independence in 1991. The best laid plans withered on the vine. The vision set out for the Tajik frontier – where industrial enterprises and state farms put people to work in the valleys while carpeting the mountainsides with trees – quickly turned to rubble.

What is made of the frontier when it is abandoned and left behind? This chapter will explore what happened when massive state intervention disappeared and an improvised commitment to enlivening the vestiges of the past took its place. The quest to make the Tajik frontier into a Soviet exhibition helped marshal immense resources and investments. However, Tajikistan's foreboding terrain presented challenges to planners seeking to enhance its ecological productivity. Five-year plans promised vast new tracts of forests girded by dams, irrigation canals and roadways – but the landscape failed to yield to the planners' will. As a result, bureaucrats promulgated disparaging views of local residents, blaming them for destroying natural resources and stealing from state farms. But when Soviet administrators abandoned the frontier following Tajikistan's independence and the outbreak of war, the projects they left behind were reimagined by Adham and his neighbours. The landscape of opportunity had been remade. The Tajik frontier's peripheral location left the region devoid of jobs and resources but afforded a favourable climate for growing fruit orchards without a meddling state. Residents reworked the ruins of past interventions to nurture new assemblages that allowed them to mitigate the remoteness of life on the margins.

Fruit varieties and livelihood strategies were grafted onto Soviet roots and rubble, ushering in unanticipated human–vegetal relations that stabilized a new way of life. Gilles Deleuze and Felix Guattari's notion of assemblages is instructive for understanding a wide variety of social phenomena in which wholes arise from heterogeneous parts. Since an 'assemblage continually passes into other assemblages', its multiple parts interact without ever producing a seamless, organismic whole (Deleuze and Guattari, 1987:325). Therefore, an assemblage helps elucidate processes of becoming rather than delineating an end form (Biehl and Locke, 2010) – helping to mark ongoing projects that engage their constitutive elements. As a result, assemblages place an emphasis on practices (Ong and Collier, 2005:17), ensuring that the resulting efforts are not 'an aggregation of the components' own properties but of the actual exercise of their capacities' (DeLanda, 2006:11). The resulting entities are always open to degradation and dispersal, held together by the commitments and affinities of their constitutive elements.

Conditions along the Tajik frontier amplified the sense of uncertainty, instability and 'chaos' that plagued post-Soviet states writ large (Nazpary, 2002), leading residents to revive frontier ruins through imaginative and unanticipated grafting practices. These ruins were the remainders of a Soviet legacy that sought to mitigate spatial distance through massive infrastructural and ecological interventions. The Pamirs benefited from Moscow's provisioning (*Moskovskoye obespecheniye*), a policy which sought to create 'centres' throughout the periphery (Reeves, 2014). The Soviet state shared access to the benefits that were supposedly limited to the core by dispersing coveted foodstuffs, scarce consumer goods and technical machinery to remote regions with critical resources (such as quartz, necessary for rocketry, in the case of Tajikistan). This legacy of robust frontier engagement, followed by shockingly abrupt abandonment occasioned by the Soviet Union's demise, forced the residents of the Pamirs to turn to new horticultural practices and labour arrangements. Residents reimagined their post-Soviet relations with the state by reworking the land in the absence of Moscow's provisioning. People brought emergent social and ecological dynamics to bear on the practices and experiences of life in a remote region.

As Adham pointed out to me outside his shed, 'we never expected to do any of these jobs ourselves. There were specialists in charge of every aspect of life, but now it's up to our hands and our will.' In place of a complex system of planned distribution, Adham chose to store apples in his makeshift cellar in the hopes of fetching a higher price in the depths of winter. The orchards that he inherited required new scions to produce more profitable, resilient, and travel-worthy fruit. These improvised techniques came to replace the discarded schemes intended to rationally mechanize collectivized landscapes. The fruits of residents' labours connected them to distant state services and far-flung family members, helping to mitigate a sense of spatial dislocation when the frontier lost its vaunted status. With no way to revert to the extractive industries and collectivized farms that had once bolstered the region, residents reinvented marginal spaces as zones of horticultural promise. Maintaining and improving the landscape of a disregarded frontier required an invigorated interest in the land, encouraging residents to breathe new life into the arboreal, industrial, and ecological remnants of the Soviet legacy.

Prior to Soviet intervention, Central Asia enjoyed a long tradition of pomiculture (fruit tree cultivation). The origins of the apple tree have been traced to the mountains of southern Kazakhstan. Cultivation was common for centuries, but not organized for large-scale production.

Under Soviet rule, the landscape became a testing ground for introducing rationalized and mechanized forms of industrial resource management. Agronomists on the Communist payroll used most of the arable land for tobacco, potatoes and, primarily, fodder for livestock (Nigmatov, 1977). When the state turned its attention to orchards, they were primarily seen as a means of afforestation. The poorly yielding fruit trees that resulted were brushed aside by planners who privileged the number of hectares planted over the quality of fruit yielded. Furthermore, managers of collectivized agriculture saw the local population as the cause of deforestation and a burden on the land. Smallholder horticulture was discouraged and rarely practised. On the eve of socialism's demise, not a single household produced orchards for sale. However, a quarter-century after the Soviet Union's collapse, fruit trees were flourishing. Through grafting practices, frontier landscapes were remade.

Fruit trees perfectly highlight the degradation of assemblages where schemes, hopes, and branches are liable to break down and wither. Assemblages offer an expansive conceptual tool to make sense of multiple scales, inputs, and histories. How can foreign apple cultivars, painstakingly-repaired tractors, and decades-old irrigation canals work together to produce a hybrid tree? The concept of assemblages offers a way for historical legacies, geographic conditions and committed actors at various scales to come into relief within identifiable composites. Assemblages help to corral the multiple components of lived reality into working wholes. But it is the efficacy of their enactment, adding up to more than just the sum of its parts, that distinguishes an assemblage from any other agglomeration of things (Rubinov, 2014). In so far as it enables actions, inhibits growth, or changes the underlying conditions, an assemblage is an active entity that engages the people, places, and things that constitute it. But once those commitments fail or diminish, then the vital assemblage as a work-in-progress ceases to exist, allowing its constituent elements to be used elsewhere or recede into futility. At the margins of the state, productive assemblages often rely on the detritus of empires and the ruins of past regimes. There is often nothing else at hand (Figure 5.1).

The Pamirs: A Showcase Frontier

The Pamir region had been a crossroads of Silk Road routes so foreboding that the region still retains its lasting moniker: the roof of the world. The troops of the Russian Imperial Army had taken over the region in the 1880s but had relied primarily on local khanates as vassals to manage the region. With an eye on the conquest of neighbouring

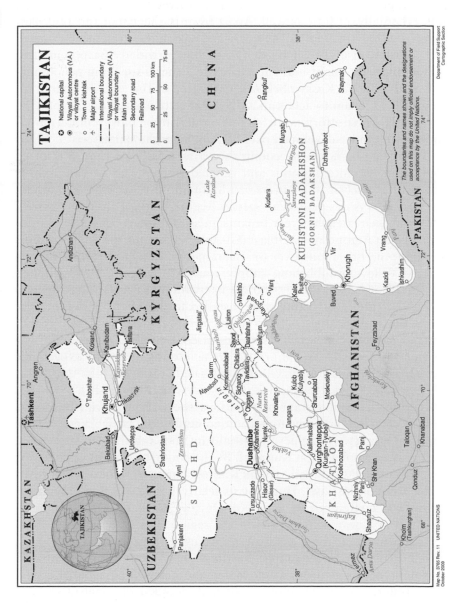

Figure 5.1 Map of Tajikistan. *Source:* Map No. 3765 Rev.11 October 2009, United Nations Organization. Reproduced with permission.

countries, leaders in Moscow decided to grant Tajikistan, then a part of Uzbekistan, the status of a full Soviet Republic in 1925. Soviet Tajiks would be promoted and their culture celebrated in the hopes of attracting their Persian-speaking neighbours across the border in Afghanistan and, eventually, in Iran as well. Apart from the fact that the Tajik language is closely related to Farsi, it was Tajikistan's strategically important location in a crucial border zone that provided 'the potential to showcase Soviet might and supremacy to Persian-speaking neighbours and their allies' (Kassymbekova, 2011: 362). The steady march of progress was meant to act as an advertisement to the nations beyond the Soviet pale. These efforts were felt profoundly in the most remote, mountainous and under-developed part of Tajikistan: the Pamirs.

Comprising 46% of the country's territory but less than 5% of its population, the mountainous and isolated eastern half of Tajikistan is known alternately as the Pamirs or Mountainous (Gorniy) Badakhshan. The region was designated a 'Special Region of the Pamir' in January 1925 and then later that year became an autonomous province: Gorno-Badakhshan Autonomous Oblast (GBAO). Soon after, the Soviet authorities built the first roads over the main mountain passes connecting the isolated valleys, establishing supply routes to urban centres (though these roads were still inaccessible for many months each winter). Several hydro-electric power plants were installed in the 1940s and electricity connections made their way to nearly every village in GBAO. Trucks brought nearly inexhaustible supplies of coal, lumber, diesel and other essentials. Medical care improved tremendously and life expectancy soared. The population of GBAO ballooned from less than 25,000 in 1917 to 170,000 in 1997 and 220,000 by 2008 (Bliss, 2010).

Apart from tangible state interventions, social life was transformed as well. Women left the home to become comrades and co-workers. Nearly universal literacy was achieved (initially in Tajik and then, in later years, with greater emphasis on Russian). By the end of the twentieth century, residents of GBAO's small capital, Khorog, began to think of their town as the 'Paris of the East'. They saw its relatively liberal values, abundant wares, and contemporary styles of dress as markers of a cosmopolitanism that elevated the status of the region. Even small villages dispersed throughout the glacial-carved valleys enjoyed movie screenings, travelling performance troupes, bustling houses of culture, and ubiquitous radio access. A formerly cloistered and inaccessible frontier region was infrastructurally, technologically and ideologically opened up.

For those who had witnessed the transformation in their lifetimes, the changes were hard to fathom. Murobo, Adham's father, had taken a fall that left him hobbled several years into his retirement. Over the course of 16 months of fieldwork spanning 2012–2017, I often found him seated in a chair outside his yard when the weather was fair. He would reflect on the transformations he had witnessed: 'We may be trapped between mountains, far from everything, but this is a special place.' The sense of being trapped made him bitter, yet Murobo still appreciated the challenges that had been overcome. 'When I went to school [in the 1940s], my family could not afford pants for me. I sat in a freezing schoolhouse with only my *chapan* (overcoat). Then the Soviet Union brought in roads, factories, and electricity. Our life was transformed. No one was cold or hungry and all the farmland was put to use.' Murobo was aware of the limits of the Soviet experiment, lamenting the corruption, misallocation of resources, and the planning mistakes. But he could not help but appreciate the significant changes that had transformed the region. 'The mountains and the river are still here but almost everything else has changed. So now we have to grow our orchards and build our lives on top of what they left behind.'

The changes that transformed Tajikistan not only altered the rocky landscape but underpinned an imaginative effort bound to far-reaching visions of what the frontier represented. The high-level functionaries in Moscow who signed off on developing the region, the engineers and managers who moved there to oversee it, and the local residents who manned the enterprise were not simply trying to reach production quotas or afforestation targets, however important those indicators may have been. Beyond the metrics, Tajikistan was a site for generating an imagined ideal, where people like Murobo could be touted as transformed Soviet subjects for the benefit of domestic and international audiences. Overcoming the physical limitations of the Pamir mountains provided evidence of the Soviet Union's dominance over material as well as ideological realms. However, in thrall to visions of industrial grandeur, the work of creating a Soviet idyll was underwritten by immense structures of concrete, asphalt, and steel alongside sweeping transformations to water, plants, and soil. When the vision of the imagined frontier receded, the profound changes wrought upon the material and built environment persisted.

Reviving Frontier Ruins

Ruins and damaged landscapes are important spaces for social generation, economic opportunism, and emergent political formations. As Ann Stoler highlights, 'To think with ruins of empire is to emphasize less the artefacts of empire as dead matter or remnants of a defunct regime than to attend to their reappropriations, neglect and strategic and active positioning within the politics of the present' (2013:11). The hulking farm implements, concrete irrigation canals, telegraph and power lines, and numerous infrastructures that increased living standards throughout the Pamirs were assiduously repurposed by residents. What emerged from Tajikistan's landscape of imperial debris was a reversion to subsistence agriculture that creatively repurposed remnant artefacts for unplanned agrarian livelihoods. Tensions persisted between former managers of state cooperatives seeking to sell off equipment for personal gain and locals who needed those tractors, fields, and pumps to make productive use of their land. Strategic control of objects became a means of enrichment for some and survival for others.

The derelict resources of a forgotten frontier were a source of tension but also possibility. Anna Tsing situates the frontier as a malleable yet powerful symbol that draws actors into intimate engagement with a lively landscape. She goes on to note, however, that 'not everyone is caught up in frontier schemes. The frontier could move on, and something else could happen in its place. The forest might regenerate' (Tsing, 2003:5103). In the absence of a developmental state, residents had to grow something new out of the rubble of past projects (see also Middleton, Lentz and Swanson, this volume). Unlike the Indonesian tropics, where Tsing observed a degraded and rapacious frontier complex, no get-rich-quick schemes made their way to the Tajik hinterlands. 'Something else' did happen. Residents of the Pamirs, observing the evacuation of their vaunted frontier status, resolved to become rooted to the land. Forests were, in fact, regenerating. Orchards were growing at significant rates, helping to improve ecological conditions in the face of widespread labour migration (Rubinov, 2016). Though it required overcoming a variety of obstacles and impediments, residents of the former frontier began to invest energy in growing orchards and improving the agricultural output of decollectivized Soviet lands. The following section will explore the legacy of Soviet land management that laid the groundwork for the human–vegetal assemblages which became so crucial to residents of the Pamir mountains.

Rationalizing Forest Margins

At the outset of Soviet resource management in Central Asia, grand designs were hatched. Following the October Revolution in 1917, the Russian Empire's holdings in Central Asia were lumped together into one large entity called the Turkestan Socialist Federative Republic. The furrowed knot of land in Turkestan's southeast became the Tajik SSR – which was 93% mountainous (where the elevation was above 1,000 m). While low-lying territories throughout Central Asia were geared toward cotton cultivation, the mountains were the repository for the glacial-fed rivers that slaked the thirst of the desert's cotton boom. To help maintain the stability of the water supply, one of the key concerns for the Tajik SSR became forest management.

State planners intended to cover their territory with forests to improve their water retention and reduce susceptibility to erosion. 'Not only should trees be planted on mountains but also on all possible locations – along all canals, roads, and river plains. Especially forest-free territories of the republic can, in a few years, be completely forested (*oblesennimi*) in the next few years!' (Botanical Institute, 1948: 7). This was a risible goal considering the steep, foreboding terrain lacking year-round precipitation favourable for forest growth. Instead, Central Asia receives most of its rainfall in the colder months, creating significant snow pack at higher elevations but little rainfall during the warmer growing season. Despite these impediments, Soviet bureaucrats were nothing if not optimistic of their capabilities.

Foresters in Central Asia had to both improve landscape ecology and produce desirable crops for Republic-wide consumption. However, forestry agencies were measured by their success in acreage, rather than quality of yield. Mechanization and rationalization worked best when plantings were uniform and simple. Considering the problem in 1980, forestry scholars noted that, 'all plantings of valued forests, fruit and nut trees, counted as if they were for defensive afforestation' (Vihrestok and Padalko, 1983: 38). These directives were set out by the Communist Party of Tajikistan, which required that foresters 'not only establish rain-fed orchards and vineyards, but create a complete expansion of rain-fed plantations on an industrial scale' (1983: 38). This strategy privileged quantity over quality. As a result, much of Central Asia's forests were envisioned as industrial fruit plantations, geared toward covering the landscape with vast swathes of timber rather than high yields of flavourful fruit. Diversification and intensification of horticulture was limited to meagre household plots adjoining homes (Rowe, 2009).

The drive to standardize afforestation efforts was led by one of the key concepts of Soviet ecological intervention: rational management. The Soviet resource management paradigm centred on enhanced productivity through rationality. Forest economic studies sought to increase forest productivity through 'complete rationalization' of the study and conduct of forestry (Yusupov, 1966: 220). Research institutes set plans, created priorities and provided biological materials to afforest Eurasian territories. Rationality was a core principle in achieving these goals because it tried to bridge the divide between plans and reality, seeking to make the unpredictability of nature a matter of knowledge and control. Rationality could not overcome the poor productivity of Central Asian territories, hampered by inadequate rainfall and rocky, meagre soils. Significant forests never took root in Soviet Central Asia.

Therefore, in the face of consistent failure in the domain of Central Asian forestry, Soviet planners sought a villain to explain away the limitations of modernizing schemes. In the view of resource managers, there was one central culprit in this failure of local population: 'the irrational actions of people, led to the destruction of a balance in nature' (Vihrestok and Padalko, 1983:25). Irrationality was the bane of foresters and had to be stamped out. Soviet engineers and planners argued that local people damaged ecological conditions, precipitating the need for rational interventions and constant vigilance to assure their success. Such was the case with Soviet forestry policy. It rested on an overt distrust for local forest users. 'In our view, there is no real doubt that at one time the forest cover in the mountains of Central Asia was much greater. The best forests are those that are far from settled areas … from the beginning fire and axe, then livestock and plough, turned forest into bare, lifeless spaces' (Kocherga, 1966:29). Soviet land use policy not only sought to administratively distance individuals from land management but also created a moral lens portraying self-motivated users as damaging and deleterious to existing resources. These views engendered a distrust of residents, which continued to pervade state forestry policy following independence. Consequently, local residents maintained ambivalent and at times antagonistic relationships with state resource managers. These tensions increased in the post-Soviet period as local users became increasingly dependent on formerly state-run lands for their livelihoods.

Lacking Experts but Not Effort

Despite the growing importance of horticulture in Tajikistan, hardly any of its practitioners had been trained in its execution nor enjoyed a legacy of horticultural tradition to turn to. The fruits of agricultural labour were the product of improvisation. The few remaining Soviet experts in the country after the civil war did not see promise in these improvised horticultural efforts. The scientists who stayed on at research institutes continued to see residents' efforts through the distrustful lens of bygone state projects. I experienced this unequivocal dismissal of local capacity when I spoke with Lyudmila, a Russian woman with large, thick-framed glasses who wielded an assertive chop to punctuate her sentences. I found her sitting at a conspicuously empty desk on the fourth floor of the Institute for Pomiculture in the capital, Dushanbe. She explained that she was a leading expert on pear tree cultivation but had not conducted new research in nearly 30 years. After taking stock of my questions about local horticultural capacity, she leaned back in her chair and responded:

> It's nice that you've taken an interest in horticulture here in Tajikistan as it has become quite crucial for people. But the standards that we once had, which were quite high, have fallen dramatically. Important advances were made in our Institute and large, impressive orchards were planted across the country. But it's all been abandoned to the local people now. What skills can you expect them to have! Really, it's a miracle they can grow anything at all.

Lyudmila was correct in stating that horticulturalists operated without significant support. Farmer Associations had been established in the 2000s to assist tenants holding newly-privatized land in the absence of previously well-funded collectivized state farms. These Associations were meant to have agronomists on staff to disseminate information to people forced into subsistence agriculture. In fact, there were hardly any agronomists to fill those positions. However, Lyudmila's dismissive view of local farmers was not grounded in reality. Based on my research with over a hundred household farmers, the government's absence did not preclude their success. Despite disparaging state experts harbouring Soviet perspectives, grafters and farmers did not let their marginalization stop them from improvising and improving the landscape they relied upon.

Tajik citizens did not look to the state to supplant the absence of Soviet skills, investment, and ideological commitment. 'The president is not such a bad guy actually, he's just surrounded by corrupt people', a clear-eyed retiree in the Pamirs told me. 'He realizes that his government is weak. Not like what it was in Soviet times. So he let the people build things up for themselves by giving everyone their own land. This was the right decision. Whatever we want this country to be, it's up to us now. He has put the responsibility in our hands.' Tajik citizens had come to recognize that, regardless of promises, the only gains to be found would be those wrought from the land that had been handed over to them. The proliferation of horticulture in Central Asia did not owe its success to experts but to skills of necessity acquired by happenstance.

This was not the vision Soviet planners had set out. Soviet foresters had sought to mechanize the mountainsides and organize ecological resources according to 'rational use'. In the process, they had decimated the region's endemic cultivation practices and traditional knowledge. Before 1938, when Soviet involvement in the Pamir region was still limited, the harvest was estimated at 40,000 tons annually, including apricots, apples, mulberries, grapes, etc. (Bliss, 2010:286). By the beginning of the 1990s it had dropped to only 5,000 tons. 'The blame lies mainly with the political planners, who decided to slash the number of orchards in order to replace them mainly with fodder crops' (2010:286). Throughout Soviet Central Asia, extensive areas of historically widespread orchards were cleared for cotton production or other cash crops, such as tobacco. These policies led households to dedicate a large part of their kitchen gardens to fruit production (Rowe, 2009). Century old walnut trees only survived through the Soviet period in 'small clumps kept by tree lovers' (Kasimov, 1989:58), although a few sizable government plantations were distributed to residents in Kyrgyzstan. But these skills were not put to use by Soviet planners. Instead, unexpected assemblages of caretakers and trees flourished in the cracks of the now-defunct Soviet edifice. Residents directed their subsistence energies into strategies that bore fruit.

Frontier Schemes: Regenerating (Post-)Soviet Orchards

The government of the Republic of Tajikistan lacked the means to continue most Soviet-era policies, particularly those dealing with the environment. As a result, natural resource management devolved to local residents. However, the state still technically owned all the land in the country. Land tenure reforms distributed land through 99-year leases which could be transferred to other citizens but not owned outright. Therefore, apart from the 2.5 million hectare Tajik National Park at the heart of the Pamirs, the largest land holder in the Pamirs was the State Forestry Agency (*leskhoz*). *Leskhoz* territories began at the edge of village boundaries and continued up along barren swathes of mountainside sprinkled with the sparse juniper stands in upper reaches. Relatively level plots accessible to irrigation had been planted with (poorly producing) orchards during the Soviet period. After the Soviet Union's demise, most of these state orchards had been relinquished to households living close to the plots, determined primarily by whomever wanted the underperforming orchards. In making the choice, residents had to choose between one of two bad options (Figure 5.2).

They could try to modify *leskhoz* territories, which had potentially advantageous irrigation systems and land terracing, but required grafting or removing old trees and establishing fences to keep out livestock. Plus, whatever outputs they could eke out might have to be shared with the forestry agency in dubious profit-sharing arrangements. The other alternative was to start from scratch on lands that had been handed over

Figure 5.2 A farmer takes pride in the trees which he grafted with more profitable apple varieties in an orchard that he inherited from a defunct Soviet collectivized farm in the Pamir mountains of Tajikistan. Photo: Igor Rubinov.

from the decollectivized state farms. 'We have to grow something useful from the land even if we don't really know how', a neighbour explained to me as he watered a row of apple tree saplings on recently privatized land. 'Whether it is *leskhoz* land or our land, it's still all owned by Rakhmon [the president]!' A few enterprising landowners turned the *leskhoz* plots, overrun by weeds and shrubs, into flourishing tracts. They chose to make the investment of time, money and effort to graft newer, more profitable varieties onto the gnarled trees. However, it was more common for energy to be invested into the land they had recently secured through land tenure reforms, where improvements might yield dividends for decades to come.

Whether on state forest lands or on their own plots, my interlocutors regularly took hybrid varieties of tree shoots and conjoined them to existing rootstocks. Over the last few years, these hybridized trees became crucial agricultural assets for the families I lived with in the Pamir mountains of Tajikistan. Given widespread emigration from rural homesteads to find work in Russia, there was a need to keep families connected. Trees helped families stay rooted at the margins of a former frontier. Nancy Peluso outlines a similar investiture of effort and commitment in Indonesia. People came to appreciate the value of durian fruit trees as urban demand for them grew, to the extent that 'economic trees now dominate the landscape' (Peluso, 1996:517). As families changed locations according to the dictates of swidden agriculture, long-living fruit trees marking past homesteads became an important component of the social, economic, and ecological landscape. People related to the landscape through the meaning and value imputed to these increasingly important trees.

In Tajikistan, economic trees were not carved out of a vast forested landscape but became part of fledging forest zones surrounding homes in rocky mountain valleys. However, like Indonesia's durian trees, Tajik orchards repurposed past infrastructures and marked ecological investments across newly vital terrains. What had once been state-run forests producing bitter fruits had become the material basis for growing delicious and profitable varietals. As marketization penetrated rural spaces, in Indonesia and Tajikistan, people began to take note of the particular properties of trees, seeing them as foundational cornerstones of new land use strategies. When sons came back from labour migration in Russia, they recognized the acute absence of work and economic opportunities. While a few of them took off again, many were encouraged to stay because of the promise of making a living with orchards. It was not that all other activities ceased, but rather that economic trees offered a way to rise above subsistence to try to make a profit and overcome the limitations presented by a marginal and remote region.

Fruitful Connections

A variety of factors encouraged horticulture in Tajikistan. The trees themselves ameliorated erosion, beautified and fortified the harsh landscape, and provided an inviting environment for life. Apart from the forested landscapes that trees generated, the most significant outcome was their fruits. Horticulture turned soil, water and air into things that could be eaten. These fruits formed the material components of transactional and social exchanges in the valley (and beyond). In so doing, they extended the efficacy and spatial reach of the human–vegetal assemblages that had come to dominate the neglected frontier. Fruits travelled beyond the boundaries of the emplaced nucleus, composed of rooted families and fields, reaching out to needed resources and connections. Whether bolstering health, assets, or status, the fruits of sylvan toil provided tangible links to make new lives amidst the deteriorating frontier.

Amon was one of the most genial neighbours in the village. He was genuinely upset every day that I did not set out to pay him a visit. One spring day, I did just that, as we sat down with a jar of honey with our tea and bread. He lamented that not many apples were coming in that year but the cucumbers and potatoes were enjoying the strange weather. The unusual weather was also giving his heart some trouble. However, Amon would have to wait till the autumn for a checkup. He had a disability pension from a bullet that grazed his pancreas during the war – affording him free treatment once a year. I asked Amon why he waited to visit the hospital in the fall if his pain usually flared up in the spring. He smiled, 'I always take a nice box of apples, big ones', he held up his hands as though grasping a chalice, 'and some walnuts'. Even though his medical care was technically free, Amon needed to supplement it with gifts of fruit.

When Amon sought to sell most of his fruit harvest, he relied on someone from outside the region to bring his goods to market. The uncertainties involved in trucking perishable wares over a dreadful road with an unreliable vehicle, not to mention negotiating a good deal with a retail buyer, were disincentive enough. Amon simply waited for a wholesaler buyer to arrive at his doorstep and offer to take his dried fruits, pears, walnuts and apples. He would have more leverage with buyers if he could warehouse the goods himself until the best time to sell. That was the reason Adham had been upgrading his backyard shed, hoping to keep his own apples later in the year, so that they might fetch a higher price. New infrastructures were emerging to buttress evolving livelihood strategies. These fruit management techniques underscored the importance of trees but also collapsed the distance of the frontier from urban centres. As Adham explained to me outside his shed, 'I won't even sell a lot of these apples. I need to send some to my son in the capital, save a few sacks for my niece's wedding, and maybe try to get my youngest son transferred to a different military base with a few well-timed boxes to the officer in charge.' If fruits and nuts could be sent to market, an army base, or to a doctor's office, then parts of the assemblage could circulate on behalf of residents – providing them with access to social services and economic opportunities that were woefully absent at a neglected frontier.

The horticultural assemblage that emerged out of the ruins of Soviet infrastructures warped the sense of remoteness that defined the frontier. The individuals who had invested their labour and capital into orchards were able to use these horticultural assets as extensions of their own labours and household connections. As the environmental and infrastructural legacies gave way to entrepreneurial imperatives, the frontier became more proximate as fruits collapsed the distance to urban centres and migrants abroad. Though growing increasingly more connected, life in the Pamirs was still a far cry from the feeling of contiguity that had been made possible by Soviet provisioning and ideological commitment. In its stead, people took the rubble of Soviet projects, enlivened them with their own tenuous commitments and used the outputs to overcome the separation and inaccessibility that still afflicted the frontier.

Conclusion

Growing desirable cultivars required a commitment to frontier landscapes through a close attunement to ecological conditions and improvisational techniques. Grafting hybrid forms enabled the remnants of the Soviet project to be appropriated for the prosaic needs of post-Soviet residents. My informants had to overcome the limitations imposed by frontier-making endeavours of the past along with the moral vilification those efforts smuggled into the present. Aspiring horticulturalists learned from neighbours, paid attention to trees, and experimented on the edges of fields. The unintended assemblages bound people to plants as they struggled to remake frontier landscapes that had once been freighted with grand ideological visions. The ruins of those bygone schemes had become the substrate for articulating new visions and forging new livelihoods. The resulting harvests allowed residents of the Pamirs to nourish themselves, circulate the fruits of their labours, and create new opportunities along a neglected frontier.

Scholarship in science studies and anthropology has prioritized the technical and expert-driven connections that sustain and define assemblages (Latour, 2007; Ong and Collier, 2005; Li, 2007b). This chapter has highlighted the quotidian and untrained appropriation of material forms as a mode through which landscapes are remade, ideas contextualized, and vegetation grown. While the interventions of the Soviet state were expert driven, the pragmatic appropriation of their remnant materials was brought about by improvised human–vegetal collaborations that continue to be reworked. As Tania Li (2007b) has noted, practices of assemblage are often organized into technical interventions in response to failures and omissions. I argue that such iterative work is not only done by state planners and development professionals but by local residents, who repurpose ruins of the past to create possibilities for the present and lay the groundwork for future endeavours.

A profusion of people and plants capitalized on limited interventions to creatively rework marginal landscapes. Where rational plans and mechanized orchard rows proved fruitless, flourishing collaborations emerged from committed investment in money-making trees and untested scions. In the process, the environment was remade, as people repurposed the debris of imperial materializations to grow forests where the state had receded. Such a terrain presents innumerable challenges, while also opening up fissures allowing for something new to grow.

Note

1   All names in the text are pseudonyms.

6

Patterns of Naturecultures

Political Economy and the Spatial Distribution of Salmon Populations in Hokkaido, Japan

Heather Anne Swanson

Introduction

Frontier processes are known for the massive redistributions they cause. Through resource extraction and the making of wastelands, their ability to shunt wealth from peripheries to centres has been well documented by scholars (Wallerstein, 1974; Wolf, 1982; Tsing, 2005; Massey, 2007). How, this chapter asks, do such accumulation practices redistribute not only capital, but also *animal and plant species* in ways that have ecological and even evolutionary consequences?

Analytically, this chapter brings together political ecology's insights on natural resource extraction with the growing attention to biological worlds within social science conversations. To this end, it draws insights from fields such as multispecies ethnography and environmental history. Its goal is to concretely demonstrate how political economy is critically important not only for understanding contests over resources, but also for understanding changes in the ecological worlds of which we are a part.[1]

It does so by focusing on the redistribution of Pacific salmon from the perspective of Hokkaido, Japan, an island that became a Japanese settler colony in 1869. The central resource that drove Japanese interest in the island, salmon – which migrate between freshwater streams and the ocean – take us directly into questions of frontier-making both on land and at sea. Their industrial harvest and hatchery production are part of

Frontier Assemblages: The Emergent Politics of Resource Frontiers in Asia, First Edition. Edited by Jason Cons and Michael Eilenberg.
© 2019 John Wiley & Sons Ltd. Published 2019 by John Wiley & Sons Ltd.

a complex multi-scalar redistribution of fish bodies in the North Pacific – one that has had rippling effects across economies and ecologies, with major consequences for both human and nonhuman livelihoods (Figure 6.1).

In contrast to many classic stories of frontier exploitation, the case of Hokkaido salmon does not end in catastrophic decline. While eastern North Pacific frontiers, such as those of British Columbia, Washington, Oregon, and California, have experienced massive salmon losses, including stock extinctions and Endangered Species Act listings, Hokkaido's salmon populations have *dramatically increased* since the 1970s. Today, Hokkaido has more salmon than at any point in recorded history. Yet the increases in salmon in Hokkaido have been geographically uneven, leaving some watersheds and fishing communities with a glut of fish and others with few. How, this chapter asks, have the intersection of particular practices of resource use and unexpected quirks of salmon biology produced these patchy distributions of salmon? What are the consequences of the new locations and migrations of large numbers of fish for both fishing communities and ecological relations?

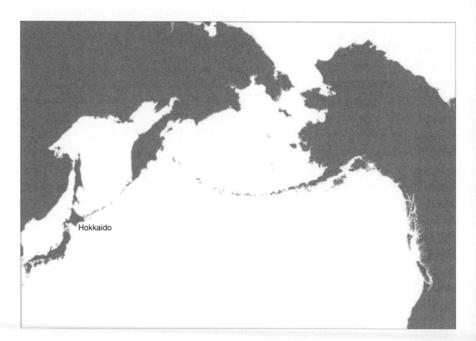

Figure 6.1   Map showing Hokkaido and the North Pacific. *Source*: 'World Continents' data set from Esri. This map was created using ArcGIS® software by Esri. ArcGIS and ArcMap™ are the intellectual property of Esri and are used herein under licence. Copyright © Esri. All rights reserved. For more information about Esri® software, please visit www.esri.com.

Overall, through its exploration of these questions, this chapter argues for more thorough engagement between this volume's notion of 'frontier assemblages' and the biological sciences by demonstrating how seemingly 'biological' processes, such as shifts in fish population structures, are profoundly shaped by modes of political economy and frontier-making.

Marginal Frontiers?

The Japanese salmon frontier that this chapter explores is doubly marginal – first, because it focuses on Hokkaido, a place not often invoked in such conversations, and second because it includes aquatic spaces, rather than strictly terrestrial ones. By presenting this case, I hope to draw attention to these edges of scholarship on frontiers, borders, and boundaries.

To call a place a 'frontier' is always a comparative move, one that draws it into dialogue with others that have been given the same label. English-language scholars studying Japan have been divided on whether or not to apply it to its northernmost main island. Although he draws extensively on the intellectual tradition of New Western historians such as Cronon (1983, 1991) and White (1995), Brett Walker, an environmental historian of Japan, has written about his ambivalence about conceptualizing Hokkaido as a 'frontier'. Walker purposefully does not use the term because he feels that its connotative baggage obscures more than it reveals, particularly in regard to the subjugation of the Ainu and the ecological transformations of their homeland, which should not be assumed to be identical to those of the American Indians and the America West (2001: 5–6). David Howell, an economic historian of Japan, similarly objects to calling Hokkaido a frontier as he thinks it 'distorts the process by which the island and its people were absorbed' (1995: 17). In contrast, historians Tessa Morris-Suzuki (1994, 1996, 1999) and Bruce Batten (2003) have argued for attention to frontier-making in Hokkaido precisely because it brings Hokkaido into conversations about comparative nation-making. While they acknowledge that the frontier is always a fraught concept, they argue that it is too useful for understanding Japanese nation-state formation to ignore. I mention these debates because they are crucial to thinking about the geographies of scholarly conversations about Asian frontiers and the marginalization of certain places within them. One of the aims of this chapter, then, is to assert a place for Hokkaido within mainstream work on frontiers.

Beyond questions of its utility as an analytical category for scholarship on Hokkaido, the frontier is clearly a critically important emic term: since the mid-nineteenth century, Japanese government officials, foreign experts, tourist promoters, and authors have consistently and explicitly conceptualized Hokkaido as a 'Japanese frontier' (Mason, 2012; Mason and Lee, 2012). Today, even on a brief trip to Hokkaido, one cannot miss its frontier-ness. As one American travel writer put it:

> In many ways, Hokkaido is the least 'Japanese' of all the main islands. It's Texas and Alaska rolled into one. It's Siberia. Switzerland. The last frontier and the end of Japan. It was not formally colonized until after the Meji Reformation of 1868, and even then it wasn't completely opened up by settlers until the 1880s – at about the same time that the American Wild West was at its peak and Doc Holliday was blasting away at the OK Corral. Hokkaido even *looks* like the American West. (Ferguson, 1998: 365)

The uncanny echoes that this author senses are the result of over 150 years of active work to make Hokkaido into a frontier, often drawing on the American West as a model. Japanese government officials specifically sought American advisors to be their guides in settling Hokkaido, which had long belonged to the Ainu indigenous people, but which the Japanese sought to bring into the fold of their emerging nation-state (Yaguchi, 2000). For Japanese officials, the Americans, who had dealt with the Indians and tamed prairies with ploughs, seemed like ideal consultants for their efforts to 'civilize' the Ainu people and introduce intensive agriculture to the island. The Americans who answered the Japanese government's requests for advisors ranged from the sitting US Secretary of Agriculture to a host of highly capable experts in mining, agriculture, ranching, and even fish processing (Fujita, 1994; Maki, 2002).

The work of these American advisors left a distinct imprint on Hokkaido's development and landscape (Fujita, 1994). In contrast to Honshu and other islands, where small-holder rice-based agriculture dominates, Hokkaido, on average, has much larger industrial farms, that cultivate crops such as potatoes, wheat, corn, and sugar beets. Hokkaido is also Japan's dairy heartland, dotted with farms that seem straight out of Wisconsin. Such rural land use patterns are a direct consequence of nineteenth- and twentieth-century government goals to make Hokkaido a place of rational, scientific resource use. They wanted to build a new modern nation and models for Japanese modernity through the island's development. Such dreams are clearly evident in the urban planning for the city of Sapporo, Hokkaido's capital city. Laid out in the late nineteenth century, it uses an American-style grid system to form square city blocks with wide boulevards, an uncommon layout in Japan (Maki, 2002). Histories of frontier-making also permeate its institutions: Hokkaido University, its most prestigious higher education institution, was built to model American land grant colleges, which aimed at creating the cadres of citizens who would modernize and develop rural America. Today, Hokkaido University continues to list 'frontier spirit' as the first of its four basic educational philosophies.

Contemporary Hokkaido is undoubtedly less 'wild' than many Asian frontiers. It is not a place of lawlessness or state absence. Rather, it is a place of paved roads, regular tax collection, and universal basic education. Yet it is a place where frontier assemblages continue to profoundly affect the lives of the humans and nonhumans who live there. Violent primitive accumulation and boom-and-bust economies have been central to the making of Hokkaido, and they continue to haunt it. Ainu activists remind the Hokkaido government of the theft of Ainu lands, and of Japanese efforts to convert the place they call Ainu Moshiri into a space for Japanese capital accumulation. They point out the common frontier-making move of the Japanese government to portray Hokkaido as a tabula rasa prior to ethnic Japanese settlement.

In this volume's introduction, the editors draw significantly on Anna Tsing's conceptions of frontiers as 'imaginative projects' which cultivate new economic and ecological forms (2005). In Tsing's work on frontier-making in Kalimantan, Indonesia, the frontier is exceptionally dynamic – as mining, timber, and palm oil companies move in dispossession and mass forest destruction happen right before one's eyes. There is a temporality to frontier assemblages: in their early phases they are often wildly expansive; later on, they are often routinized and rendered mundane (see Rubinov, this volume). This is by no means an end to frontier formations. As Morris-Suzuki describes it, frontier development entails 'a restructuring of the relationship between humans and nature' across multiple terrains (Morris-Suzuki, 1998:25). Because such relations are material and embedded in landscapes, frontier assemblages remain in patterns of wealth and power, in waste dumps and damaged environments, and in the bodies of people, animals, and plants. These persistent changes are what scholars such as Tsing (2015) and Stoler (2013) have called 'ruination'. But sometimes, as the case of Hokkaido reminds us, ruination can take the shape of increased resource abundance – alongside depletion – as biological patterns become tied to spatially uneven accumulations of capital (see also Choi, this volume).

Hokkaido's Salmon Frontier

Although salmon have been at the centre of northern Asian indigenous trade networks for thousands of years, the industrial harvest of salmon in Hokkaido, began during the seventeenth and eighteenth centuries, when ethnic Japanese merchant-traders gradually stopped trading with the indigenous Ainu people for salmon and began establishing their own system of salmon harvesting stations with the Ainu as forced labourers (Howell, 1995; Walker, 2001). These salmon – dried or salted – travelled south by boat to Honshu, where they not only ended up on the tables of Japanese elites, but also in rice fields, where they were valuable fertilizer. Yet despite their intense exploitation of Ainu labour and the island's fish, these practices were not experienced through the categories of either 'frontier' or 'capitalism'.

Salmon canning, which began in Hokkaido in 1877, was substantially different. In the second half of the nineteenth century, it became a classic frontier industry around the North Pacific Rim, one entangled with indigenous dispossession, unbridled accumulation, and new modes of transnational trade. While political ecologists have written extensively about fisheries and marine resource exploitation, and sea tenure (Acheson, 1975; Durrenberger and Palsson, 1987; Cordell, 1989), that work has often sat outside conversations about 'frontiers' per se. While several of the chapters in this volume notably take up new domains, such as air, water, and subsurface spaces (see Zee; Choi; Günel, this volume) frontier scholarship continues to suffer from a terrestrial bias – an assumption that frontiers are found solidly on land rather than in the water or at its edges.[2] But as imaginative projects, forms of frontier-making have routinely flowed across terrestrial–aquatic boundaries.

As the American frontier booms brought more white settlers to the US West Coast, they scrambled to make their fortunes from fish along with gold, crops, and railroad development. In 1866, four men from New England – drawing on canning technology developed a half century earlier in France – established the first commercial salmon cannery near the mouth of the Columbia River along the borders of Oregon and Washington states. Canning made salmon easier to ship long-distances without spoilage, allowing fish captured along the US frontier to tap into urban East Coast as well as European markets. As was the case for the Chicago stockyards and the Argentinian beef industry, where new forms of refrigeration fostered the emergence of particular resource frontiers, canning technology played a similar role for salmon fisheries in the North Pacific (Cronon, 1991). By 1883, less than 20 years after the first experiments with salmon canning, 39 large canneries were exporting Columbia River salmon to distant markets (Lichatowich, 1999).

The salmon canning industry rapidly spread northward to British Columbia and Alaska like an aquatic gold rush (Arnold, 2008), and indeed, the industry produced similar cultures of violent masculinity and frantic accumulation. These processes did not go unnoticed in Japan. After the Meiji Restoration in 1868, when the new Japanese government aimed to colonize Hokkaido and incorporate it into the emerging Japanese nation-state, they were not only interested in sparking the development of its lands, but also its seas. As soon as they got wind of the US salmon canning industry, members of the Hokkaido Colonization Bureau, a government division, both commissioned a report on the salmon fisheries of the Columbia River by an American expert and sent members of their own Japanese staff to directly observe the booming salmon canneries on the far side of the Pacific. With the help of an American canning expert whom they wooed to Japan, the Hokkaido Colonization Bureau quickly set up Japan's first cannery near the mouth of the Ishikari River, not far from Sapporo (see Swanson 2018).

Although Japan is renowned as the world's largest per capita fish consumer, salmon canned near Sapporo was not destined for Japanese tables. The Colonization Bureau viewed salmon as an export product capable of bringing needed foreign currency into the new nation, rather than as a domestic food source, and thus, canned salmon came to travel from the canneries of Hokkaido to the docks at Yokohama to the markets of London and Paris, where they began to compete with Columbia River products. Because the price of imported Pacific salmon was less than that of meat, almost overnight, canned salmon became a staple protein source for the British working class.

The rise of the industry depended – as usual – on private companies whose accumulation was made possible due to state support. In addition to cannery R and D, the Japanese government passed and enforced laws that blocked Ainu salmon harvests in order to make more fish available for industrial canning. But the immense success of the salmon canning industry quickly began to impact the fish's populations, and by the 1890s, fish stocks in Hokkaido (as well as the Columbia River) were already beginning to show signs of decline. In the case of Hokkaido, salmon fishermen and processors began pushing the salmon frontier northward to the Kuril Islands and Kamchatka.

As part of negotiations at the end of the Russo-Japanese War, Japan received rights to establish fishing outposts on the Kamchatka peninsula, and Japanese companies rushed to set up a network of canneries to harvest these more northerly fish. While these populations were showing signs of decline as early as the 1920s, they continued to support a robust Japanese salmon canning industry until the early 1940s, when World War II halted operations. When the Japanese salmon fishing resumed in the post-war era, Japanese fishing boats – increasingly large factory ships – both returned to fishing grounds near Russia and began plying waters further afield, including ocean areas off the coasts of the US and Canada. By the 1960s, however, it was clear that Japanese abilities to exploit new oceanic salmon frontiers were likely to end as international legislation increasingly restricted high-seas salmon fisheries.

At this point, the Japanese industrial salmon industry fractured. Some companies turned to speculation in buying and selling Alaskan caught salmon; others invested in salmon farming ventures in Chile (Swanson, 2015). In short, big companies largely got out of the business of catching salmon and turned to other kinds of fisheries frontiers. Japanese salmon fishing was in ruins, but it did not die. It moved back to a place it had long ago left: Hokkaido. As large-scale salmon fishing moved northward, countless others had moved in to make what they could from Hokkaido's near-shore fisheries. Although some families were able to incorporate fisheries into larger accumulation strategies and establish small fishing fiefdoms, by the early twentieth century, Hokkaido's depleted salmon fisheries were a pretty poor bet for newcomers seeking to make a fortune. Still, second and third sons from Honshu – those who would not inherit their fathers' farms – often moved to Hokkaido to stake out a new life. While many tried their hand at farming, others took up fishing; most of these ended up in poverty, trapped in contract fisheries that left them in constant debt.

After World War II, the Japanese engaged in a process of 'sea reform' in Hokkaido to grant ownership of fishing rights to fishermen, rather than absentee owners. Yet they invested relatively little in these marginal small-scale fisheries, until the impending closure of industrial high-seas fishing. Triggered by that loss, along with larger concerns about Japan's low food self-sufficiency rate, they began to try to build a salmon industry in Hokkaido. But doing so was challenging in a place radically transformed by Hokkaido's land-based settler colonial efforts. Hokkaido's landscape had been reoriented; it had been converted from an Ainu-managed landscape of forests and wetlands to one of large-scale monocrop agriculture mixed with patches of urbanization and heavy industry. This reorientation of the landscape for particular forms of capital production proved disorientating for salmon. The draining of marshlands, construction of dikes and dams, gravel mining in rivers, pollution from agricultural runoff made Hokkaido's waterways inhospitable to spawning fish.

The Japanese government, however, aimed to bring them back through the establishment of hatchery-based sea ranching. It was an attempt to generate new resources and profits in the ruins of prior frontier projects. Hatchery technologies had existed since the late nineteenth century, but they had not previously had much success in bolstering fish numbers. Beginning in the 1970s, however, due to the experimentation of government funded researchers and hatchery technicians, hatcheries began producing successful salmon smolts – those which could be released into river mouths, would migrate in the ocean, then follow their spawning instincts back to the Hokkaido river of their birth at a high enough rate to make the process profitable.

On one hand, this process of producing young salmon in hatcheries, then turning them out to ocean pastures gradually came to work better and better. Hokkaido's salmon populations grew exponentially. In 1995, Hokkaido salmon fishermen hauled in 57 million chum salmon (Noakes et al., 1999), and their numbers have remained high ever since. On the other hand, however, sea ranching has produced troubling new economic and ecological problems – *elaborations* of frontier processes rather than their *remediation*. To examine them, I begin with an anecdote from the approximately 18 months of ethnographic fieldwork I conducted on fisheries management in Hokkaido between 2008 and 2010.[3]

New Distributions

It is a rainy late fall day, and I am sitting in the office of a Japanese fisheries scientist and manager. 'What's the biggest problem for the future of salmon management?' I ask, expecting an answer about global warming or hatchery disease outbreaks or degraded rivers. But instead, the scientist answers: 'Inequality'.

This inequality stretches our typical understandings of the term; it is a complex knot of social and biological contingencies – about flows of wealth as well as the vagaries of fish. While hatcheries returned salmon to Hokkaido, they did not return them to the same places, people, or ecosystems where they had previously been. And they did not distribute them very evenly. The fishermen who held near-shore rights profited from the new Hokkaido hatchery salmon system, but those who had laboured for high-seas companies were left with nothing. The companies picked up and left, and the vibrancy of the cities out of which they were based in far eastern Hokkaido disappeared along with them. When I visited one of these cities, I was struck by the local community's losses. With little industry left and a now declining population, the city broadcasts a soundtrack from loud-speakers at the tops of telephone poles in an eerie attempt to fill the emptiness of a commercial district with a growing number of shuttered stores.

Yet even among the near-shore fishermen, there was a major problem of inequality, and it was that which was troubling the Hokkaido fisheries manager the day I visited him. When he looks at the total number of Hokkaido salmon, everything seems fine with the statistics showing strong, stable runs. But it is the *distribution* of those salmon that the fisheries manager identifies as problem. He sketches a map on the back of a piece of paper. While salmon are returning in droves to Hokkaido's northern coast, their numbers are weak in rivers that flow into the Sea of Japan.

It has not always been this way. Japan's first salmon cannery was sited in a Japan Sea fishing town in southern Hokkaido because the area was famous for its vibrant runs. Yet today, while fishermen up north along the Okhotsk Sea coast individually net as much as $300,000 from salmon, southern Hokkaido fishermen struggle to earn even $10,000 from the fish. Such differences have major social and natural consequences. While northern fishermen spend the winter and spring playing pachinko and travelling internationally, southern fishermen spend nearly everyday out on the water increasing the harvest pressures on other fish species such as herring and flounder. When the Japanese government set out to produce salmon, it did not actively seek to produce such starkly contrasting areas of glut and absence. But neither has it taken steps to try to address it. Rather, they have taken a laissez-faire approach to the intertwined effects of capital and fish genetics.

The question of why hatchery production has become so uneven is multifactorial. Although hatcheries themselves are fairly evenly distributed around the island (as well as down the northern Honshu coast), their successes have not been. Owing to slight genetic differences, the fisheries manager tells me, the eggs of southern Hokkaido fish are slightly smaller than those in more northern rivers. While this small difference does not seem to matter when salmon spawn on their own in a stream, it appears to make a big one in their amenability to hatchery production regimes. The eggs in southern hatcheries tend to result in juvenile fish that are smaller and allegedly less agile swimmers compared to fish farther north. Because there is a strong correlation between the size of hatchery fish at release and adult return rates, southern hatchery managers try to manage the water temperature and alter feeding regimes to try make their small fry grow bigger, but their efforts haven't been very successful. The fish are smaller upon release, and they return at lower rates.[4]

But there is also a question about economic resources. Initially, Japan relied on a mix of public and private hatcheries, and in recent decades, the Japanese government privatized the hatcheries it once funded, leaving nearly all salmon production in private hands. In practice, those hands belong to the fishing cooperatives that hold the rights to harvest fish in each region. Typically, fishermen turn over about 5% of the income from their harvest to their local hatchery association, which, in turn, produces their fish. This means that wealthier fisheries can afford to produce more fish in future years, while already poorer ones struggle to do so. In this way, the initial differences in fish egg size have been further amplified. If

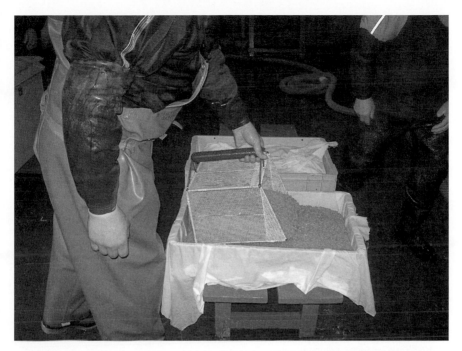

Figure 6.2 A hatchery worker prepares recently fertilized salmon eggs for transfer to an incubator. Photo: Heather Swanson.

one has a low salmon harvest, one lacks the funds to invest in the latest hatchery technology or hire the most highly trained staff in order to improve their hatchery salmon survival rates. Once started, the feedback loop between lack of funds and lack of fish can be hard to stop, and can pose a real threat to future of salmon stocks in undercapitalized regions. This is another form, but perhaps a less noticed one, of what Kaushik Sunder Rajan and others have called the 'capitalization of life' (2012). Salmon in Hokkaido are not only given a financial value and made subject to intensive harvest; their genetic futures are also bound up with flows of money and systems of value (Figure 6.2).

Prior to frontier-making in Hokkaido, its salmon spawned in a plethora of small rivers, adapting to their minute differences and developing substantial genetic diversity. Salmon have an exceptionally high degree of intra-species diversity due to their homing behaviour and its production of partially reproductive separate groups. Indeed, salmon are so adaptable and evolve so rapidly that they are changing how scientists understand processes of evolution itself (Araki et al., 2008). As salmon have tangoed with the rapidly changing geology of the Pacific Rim, they have modified their spawning times and rituals to accommodate new temperatures and flooding patterns every time they altered their range. Rather than becoming specifically adapted to relatively stable places, salmon became specialists in the art of adapting to ever-changing ones.

Salmon are also clearly sensitive to humans, too. Although data is limited, it is highly likely that the harvest strategies of indigenous people significantly shaped the evolutionary trajectories of salmon across the Pacific Rim (Colombi and Brooks, 2012). What we know for sure is that, in the twentieth century, the size of the mesh used in Japanese high-seas fishing nets caused chum salmon to evolve a smaller body size within a few decades and generations (Fukuwaka and Morita, 2008). While human influence on salmon evolution is not new, it is taking increasingly intensive forms. Hatcheries are both selecting salmon breeding partners and exerting new evolutionary pressures on eggs and young fish. A hatchery's metal egg incubator does not likely select for the same traits as would a rocky stream bottom. The same holds for juvenile fish who are now selected for their ability to grow on a pelleted diet and thrive in a tank, in contrast to previous pressures for those who were good at catching insects. Scientists have identified that salmon are so sensitive to such selective pressures, that after only a single generation in a fish hatchery setting, salmon that are returned to streams show less aptitude for spawning there than salmon who have not experienced the selective pressures of the hatchery (Araki et al., 2008).[5]

In Hokkaido, hatchery inequality has further torqued salmon populations and their genetics. When salmon stocks are dependent on hatchery production, they are also dependent on the funds to make them. This has skewed the relative balance of different salmon populations. Those whose hatcheries have comparatively little funds or those who were never enrolled in hatchery cultivation at all have shrunk, while Hokkaido's best-funded populations of salmon have come to dominate the North Pacific. This introduces new types of precarity. When global salmon market prices drop, so do the prices at which Hokkaido fishermen can sell their catch – which reduces the funds available for hatchery production. While fishing cooperatives have sub-stantial savings and insurance programs, it is not impossible to imagine a situation in which long-term price declines would reduce the number of Hokkaido salmon, with rippling effects for the ocean food webs of which they are a part. In this way, might the number of marine mammals in the Okhotsk Sea (who feed on Japanese salmon) thus rise and fall with global commodity prices? Financial precarity now begets ecological precarity.

Overall, the genetic traits of salmon who succeed in hatchery systems are vastly over-represented not only in Hokkaido, but around much of the Pacific Rim. Are these the stocks which will be best suited to changing ocean temperatures and the fluctuations of climate change? As in agriculture, too heavy a reliance on a single crop or strain is dangerous in times of fluctuation.

Conclusion: Patterns of Naturecultures

Political ecology has long focused on questions of who has control over natural resources. Here, I have tried to show how politics make their way inside ecologies, too – into the bodies of salmon. The current populations of Hokkaido salmon and their genetic diversity cannot be understood separately from the island's history of frontier-making. Hokkaido's new kinds of salmon precarity are part-and-parcel of the island's entanglement with frontier assemblages.

One of my goals, here, is to contribute to the growing body of scholarship that shows why and how political ecology and biology might benefit from more engagement with each other.[6] While political ecology pays attention to political and economic arrangements, it forgets about the ways that things like the size of fish eggs matters to patterns of inequality. Similarly, it too often forgets the profundity of how frontier assemblages permanently alter the more-than-human diversity of our worlds, even at the level of the gene. Yet, for the most part, while biologists worry about the massive changes to salmon populations, they tend to gloss them as generalized anthropogenic change, rather than viewing them as the product of specific economic and political histories.

These are longstanding critiques of political ecology and biology raised by many in and beyond such disciplines. Science and technology studies scholar Donna Haraway is among those who have urged us to try to better think biologies and politics together. In the title of this chapter, I invoke her concept of 'naturecultures' – which she writes with no space or hyphen to signal the fundamental inseparability of concepts often rendered separate (2003). My aim in doing so is to remind us of the depth and complexity of the entanglements of political economy and biology in frontier-making and life in its ruins (see also Lentz, Middleton, and Rubinov, this volume). It is a humble effort to foreground how frontier assemblages are not mere political-economic formations written onto landscapes, but are instead emergent patterns of naturecultures.

Hokkaido's salmon worlds have come to have a problematic naturalcultural pattern – one of unequal and likely unstable salmon bounty. It is a pattern whose analysis demands attention to both fish biology and political-economic histories. Political ecologists have indeed captured the ways that frontier extraction often leads to fractured ecologies, species declines, and even extinctions. What we need now is more attention to both the economies and ecologies that form in the ruins and the ways they shape each other (see Tsing, 2015). It is not only that people are forced to adopt new life ways in the midst of frontier ruination; nonhumans are, too.

In closing, I want to return to the vastly expanded sense of inequality toward which the Hokkaido fisheries manager points us. For too long, social scientists have seen inequality in strongly humanist terms – as a human problem driven by the appropriation of land, labour, and wealth of some by others. Such inequalities, however, also extend to the nonhuman worlds, where unequal patterns of human accumulation shift distributions and balances among fish stocks. While biologists have noted such changes with mounting concern, they too have not typically conceptualized them as inequality per se. Expanded notions of naturalcultural inequality, then, seem a potentially fruitful site for probing forms of scholarship and politics that might further integrate ecology and economy.

Notes

1 Overall, this effort is intellectually indebted to the work of environmental historians who have attended to relations of capital, landscape, and colonization, such as Alfred Crosby (1986), Virginia Anderson (2004), and William Cronon (1991).
2 The discussion of classic texts in the introduction to this volume illustrates this strong orientation toward land in frontier scholarship. For a scholar who explicitly challenges this, see Steinberg (2001).
3 I have also conduced short follow-up visits in 2011, 2014, and 2016.
4 There are also possible differential effects from local water temperature differences, as well as possibly from differential effects of climate change. Although all juvenile salmon migrate to the Sea of Okhotsk soon after their release from hatcheries, the water temperature and availability of food resources in the area surrounding their immediate release location are thought to have a significant impact on their survival. It seems that conditions along the Japan Sea coast are currently less favourable than those up north, leading to differential rates of juvenile survival. There is also some concern that warmer ocean temperatures could impair the return migration of some southerly adult fish. As an aside, this research occurred prior to the Fukushima Nuclear Disaster. Furthermore, for all of that event's catastrophic consequences, it has not likely had much effect on the stocks in question.
5 In another stunning example, new evidence suggests that after only 50 years, salmon populations in the Columbia River are evolving a life history pattern specifically adapted to a waterway characterized by large hydropower dams (Williams et al., 2008).
6 Examples include Peluso (1996), Boyd (2001), Prudham (2003), Fairhead and Leach (1996), and Tsing (2015), among others.

Part III

Frontier Expansions

Framing Essay
Assembling Frontier Urbanizations

K. Sivaramakrishnan

An organizing concern of this volume, in the words of the editors of the collection, is the transformation of marginal space into frontier zone. Marginality here does not signify a distinct or compact geographic location but a spatial entity shaped by relative powerlessness and ecological vulnerability. And frontier in this usage indicates both exploitation and creation; the destructive regeneration of land (in particular), but also water, and for that matter social relations, into something that reproduces the flights of fancy and desire in which a landscape's future is imagined. To better explore and explain these processes as they occur across Asia in the early twenty-first century, the authors define and deploy the term frontier assemblages. In this section the term is largely applied to processes of urbanization in frontier zones.

All contributors emphasize the contemporary distinctiveness of these processes. This is evident in the very choice of the term assemblage, described as having a short history, and its alignment with frontier, a term with a deeper and chequered past, while also choosing to discuss economic processes in terms of liberalization, speculation, and technological capacity. But what is striking, as well, is that key elements of frontier making that were significant in the work of early modern empires, and modern European colonial expansion, remain salient in the way frontiers are discussed in these chapters. Every case in this section points to the growing efficacy of the national and regional state: not only evinced in its control over territory, but also in the fostering of

Frontier Assemblages: The Emergent Politics of Resource Frontiers in Asia,
First Edition. Edited by Jason Cons and Michael Eilenberg.
© 2019 John Wiley & Sons Ltd. Published 2019 by John Wiley & Sons Ltd.

investment in innovation, service delivery, and vigorous trade relations. In that sense, a synergy that John Richards saw in the early modern world – between resources, states, and markets – remains the hallmark of these new frontier assemblages that generate novel urban forms in relatively 'remote' areas (Richards, 2003: 57).[1]

This brings me to the other part of this analytical pairing: assemblages. It is a concept that works well with the idea that frontiers are actively being made in conditions of uncertainty, unruliness, and aspiration. Many proponents of the term assemblages have seen them as open-ended formations, therefore, unlike systems or communities or even networks. What these authors are doing actually illustrates well an account of assemblages that is provided by Anna Tsing when she notes, 'patterns of unintentional coordination develop in assemblages'. She goes on to explain that 'assemblages cannot hide from capital and the state; they are sites for watching how political economy works' but not as a predictable trajectory of capital (Tsing, 2015: 23). In every case, we are reminded, then, that be it coastal China, Inner Mongolia, or Manipur in the Indian northeast, the frontier assemblage is emerging from fresh perspectives or powerful stereotypes, where the intrepid explorer can see bounty in land and this travelling gaze encounters social and political conditions which offer multiple pathways to the realization of bountiful futures.[2]

Arguably the incompleteness and unevenness of the work of colonial empires and even powerful nation-states when it came to reshaping frontier zones leaves open the possibility that the contingent account enabled by use of the term assemblages could have a longer history as well. But that said, it is important to recognize what these chapters are rightly insisting is perhaps a unique aspect of frontier assemblages in the twenty-first century. Young Rae Choi underlines the orientation toward certain ecological futures as the novelty. She shows that sustainable development in China is centred on urban transformation and rests critically on a model of constructing eco-cities on the coast where older conflicts over remembered and treasured land uses, and the historical resonances of built environments are diffused if not entirely absent, and certainly not part of hegemonic imaginations. Thus, the fishing cultures of the Tianjin area, being reclaimed for an eco-city, are memorialized in monuments or museums even as they are erased from the active landscape. This is of course a practice not unfamiliar to scholars of settler colonialism in the Americas, and can be found in the experience of many tribal and indigenous communities who were displaced from their land but conserved in cultural history as they were incorporated into national accounts of diverse origins for the modern nation in different parts of Asia.

Young Rae Choi examines the making of a land frontier on the coast, as in urbanizing China land gains importance over water for the valued future that can be imagined only through the conversion of water into land. Such a view is powered by a long-standing narrative of land scarcity in China, mostly addressing questions of arable land, but now encompassing all kinds of land. Choi makes the important distinction that coastal land reclamation in the twenty-first century responds to a prevailing conflict over agricultural (rural) land conversion to urban and industrial uses, thereby siting new urban growth on spaces that appear in the dualistic scheme as neither urban nor rural. Such a spatial logic also facilitates a specific ecological imagination where present destruction will yield the production of future greening and sustainability, once sited firmly on land. In that sense, coastal China is encompassed in a pattern of 'modelling, planning, and interpolating the future of resources and environments ... an increasing feature of contemporary environmental politics' (Mathews and Barnes, 2016: 10). But it is also part of the older project of land reclamation from the sea in the expansion of coastal cities. Such projects create artificial natures in at least two ways. One is the immense financial and engineering mobilization that creates land from water; the other is the simultaneous projective and rhetorical construction of a future reality.[3]

Max Woodworth provides another case study of urbanization in China, this time in Inner Mongolia, in which he brings together the massiveness of extractive processes in a frontier zone with the literally gigantic nature of the built and reshaped landscape that emerges through large construction projects that stabilize certain representations of the contested frontier and seek to ensure its immutability. There is much that is reminiscent of earlier resource frontiers in the account of Inner Mongolia provided by Woodworth. The region is constituted as a frontier by becoming a supply zone for raw materials and natural resources for economic development elsewhere. Such spatialization of territories within nations has often been discussed as deliberate underdevelopment or internal colonialism by its critics. When it continues across countries, of course, such a division of landscapes of extraction and production may not always predictably map on the prosperous and poor nations divide (see, for example, coal exports from large tracts of Australia or the wealthy Middle Eastern oil-exporting nations).

However, as a recent comparative analysis has shown, levels of economic development and urbanization may help less developed countries reduce the adverse effects of environmental degradation (often measured as per capita ecological footprint). But in as much as these adverse environmental effects are largely linked to trade (and export of primary products), it is the structure of the exports that may actually determine the extent of degradation (Jorgensen and Rice, 2007). Such studies can conceal the spatial distribution and intensity of ecological destruction and transformation, but they do usefully highlight the pervasive role of trade, be it within regions of countries or across nations, in shaping the frontier zones that emerge and suffer serious environmental losses.

Like the chapter on coastal China, the one on Inner Mongolia also finds the main driving force for the making of resource frontiers to be one of rapid urban development. But, as Woodworth notes, such urbanization is often about land speculation and it does not always lead to a lived urbanism – large empty buildings capture land value (always growing) and provide urban vistas which are a curiosity and can generate their own non-resident spectatorship. This finding is at some odds with the call of the volume editors to not overly weight the forces of financialization in understanding the formation of resource frontiers in the contemporary historical moment. But it is important to note, as Woodworth shows, that financialization produces its own medley of cultural representations, shifting aspirations among rich and poor, and regimes of representations around financialized resources, most importantly and evocatively – land.[4]

The frontier assemblages described and analysed in the Chinese cases are about literally extracting land or speculating on land conversion. In contrast, the case from northeastern India, Manipur, is one of creating a zone of contested sovereignty by specialization in land use. The growth of a private health sector provides both, as Duncan McDuie-Ra argues, opportunities for urban expansion and the resources to regional medical professionals, private capital, and the state government to challenge the ability of the central state to support and manage such expansion.

This is also a case which calls into account the historical record of nation building in India since the mid-twentieth century when large swaths of the northeast were brought under special administration and legal regimes in the effort to provide enhanced local autonomy in tribal districts and states.[5] Land tenure policies themselves generated a prolonged process of frontier making, exacerbated by slow rates of economic development and the domineering presence of the Indian Army for national security on international borders and in perpetual battles with militant insurgents seeking alternate futures for the region. The creation of good, reliable, private medical facilities in Imphal and Langol enables the border crossings within the wider northeast that remain heavily policed and restricted. The travel restrictions, McDuie-Ra shows, are harder to enforce when these crossings are driven by the pursuit of essential services. Thus, McDuie-Ra perceptively notes, a new resource frontier is produced. Even people from Myanmar may find their access to Manipur eased under the influence of the demand for medical services and the power of private enterprise to expand its markets.

The case studies from China and India in this section, thus, underline the value of bringing this trade-based perspective (including the trade-able nature of special and essential services) into sharp relief. When read together these chapters fruitfully inform and enrich our understanding of frontier assemblages and sharpen that analytic gaze through the case of urbanization in a series of places far from the metropolitan centres more familiar to the study of urban concentration.

Notes

1   My use of remote as a spatial signifier in these cases, follows the work of Hussain (2015).
2   In the rich literature on the making of colonial frontier zones, this point can be illustrated by a combined reading of Arnold (2006) and Burnett (2001).
3   I am indebted for this point to Andrew Toland (2017), who also usefully discusses the relation of artificial to natural in Chinese traditions. See in particular, pp. 102–103.
4   Much ink has been spilled on the question of land speculation in recent decades and the aggressive pursuit of land control for food security, but also speculative reasons, by sovereign wealth funds, large corporations, or consortia of private-equity based investors. Such targeting of land by financial markets is often discussed as an important part of land grabbing. This worldwide phenomenon has generated a considerable scholarly literature, especially in the pages of the *Journal of Peasant Studies*. For a recent review of the structural effects of the financialization of land on small holders, see Fairbairn (2014).
5   For a political discussion of some of these often violent policies of exception, see Baruah (2007); a longer duration history that recognizes the disruption of various connections and flows across the northeast after colonial rule was established, is provided by Chatterjee (2013). See also, for a situated account along these lines from Mizoram, adjoining Manipur, Pachuau and Van Schendel (2015).

7

China's Coasts, a Contested Sustainability Frontier

Young Rae Choi

Introduction

In September 2014, I attended the International Workshop on Intertidal Wetland Conservation and Management in the Yellow Sea Provinces of China hosted by the Beijing Forestry University. The workshop included over 160 participants: officials from national and provincial levels of the Chinese government, scientists, policy researchers, and representatives from international and national NGOs and project organizations concerned with the conservation of tidal flats. Discussion was fuelled by reports from numerous sites along China's eastern coasts, all indicating uncontrollable practices of land reclamation and the subsequent loss of tidal flats at an alarming rate. Local officials demanded that the central government pass stronger regulations and more effectively enforce the existing ones. Yet, it was clear that even the State Oceanic Administration, China's top-tier government apparatus in charge of governing its coastal and marine space and resources, had little power when it came to safeguarding the disappearing tidal flats. The conference participants lamented that the Chinese state's recent eagerness for sustainable development did not seem to apply to coastal territories that deserve equal preservation for social and ecological advancement. The workshop ended with a concluding declaration that illustrated the uneasiness and the urgency that were widely felt at the workshop:

> The 18th National Congress of the Communist Party of China identified eco-civilization as the national strategy. However, the implementation of

Frontier Assemblages: The Emergent Politics of Resource Frontiers in Asia,
First Edition. Edited by Jason Cons and Michael Eilenberg.
© 2019 John Wiley & Sons Ltd. Published 2019 by John Wiley & Sons Ltd.

national polices including ecological redlining, a national wetland conservation policy, and eco-compensation, which began a new era for coastal wetland conservation, never happened ... The status of reclamation and degradation of the Yellow Sea wetlands is extremely shocking and is resulting in a very critical situation. In the last decade, 2% has been lost every year. 1.4 million ha has been lost since 2005, making up 22% of the total area of wetlands.[1]

China's coasts are going through a massive socioecological transformation. Land reclamation, accused at the conference and elsewhere of being the most significant threat to the biodiversity conservation of tidal flats, is a practice that creates land by filling in coastal wetlands and shallow waters. The latest reclamation boom that began in the early 2000s, enrolling every coastal province across China, is unprecedented in its scale and speed. Between 2003 and 2005, the annual reclaimed land area surged more than fivefold, from 21.23 to 116.62 km². Since then the annual average has stayed above 100 km². Roughly speaking, China adds the land equivalent of two Manhattans every year. The total area of land that China has created post-millennium is over 1,450 km².[2]

While the insatiable desire for land is fast encroaching upon the coasts, resistance to land reclamation for environmental reasons is becoming even more difficult as the practice of reclamation is increasingly perceived as 'ecological' in twenty-first-century China. This phenomenon signifies a break from previous assessments of coastal reclamation as one of the greatest and irreversible causes of marine biodiversity and habitat degradation in the East Asian region.[3]

While it seems to stand in opposition to the now popular 'eco-' agenda of the Chinese state, reclamation projects today acquire legitimacy under the emerging paradigm of sustainability. They do this by incorporating eco-cities, renewable energy infrastructure, carefully designed non-human habitats, and nature reserves into their designs. In contemporary China, reclamation is no longer about destructive development; it is about creative production (see also Paprocki; Zee; and McDuie-Ra, this volume). Coastal land reclamation is likewise not a mere land creation project; it is about envisioning new materialities and values about what China's future should look like (see Woodworth and Günel, this volume).

This chapter examines the recent transformations of China's coasts into resource frontiers entangled with the country's broader and deeper societal transformations. Specifically, I look at how China's coastal spaces are reconfigured as repositories for land provision *and* as an experimental and an idealized platform where China's new vision for its sustainable future is projected, tested, and demonstrated (see also Paprocki, this volume). Along with others in this volume, this chapter conceptualizes China's coasts as an emergent 'frontier assemblage' that is invented through the works of multiple contingent forces and processes and is subject to new and more intensive forms of exploitation. Thinking through the notion of frontier assemblage, I draw upon Tania Li's works on assembling land as an investible resource (Li, 2014a, b) and Anna Tsing's insights on frontiers (Tsing, 2003, 2005). A resource, Li argues, does not possess an inherent quality to function as a resource. It is produced as a resource through complex assembly work of 'heterogeneous elements including material substances, technologies, discourses and practices' that render something valuable at specific moments in time and place (Li, 2014b: 589). Likewise, a resource frontier does not pre-exist but is made as a frontier through the works of material and imaginative renderings (Tsing, 2003).

Both Li and Tsing emphasize a conjunctural approach to the formation of a resource frontier. China's coasts, I argue, are situated at a conjuncture of a highly particular political economy of land and an emerging politics around sustainability as China's next model of economic and social development (see also Anderson, this volume). This conjuncture does not merely frame post-millennial coastal land-reclamation practices. It shows where China is heading against the backdrop of, on the one hand, the predicament of an urban-driven growth strategy that the country has moved toward for over two decades and, on the other, the socioecological crises that China's modern development experiences have brought in turn. The following sections of this chapter examines the way these processes render coastlines as new resource frontiers: at once producing for land itself and spatializing sustainability onto reclaimed land.

Making a frontier entails 'conjuring' certain positive imaginaries about the frontier (Roy and Ong, 2011; Li, 2014b) and rendering the place as backward, underdeveloped, messy, and even empty (see Cons and Eilenberg, this volume). Gavin Bridge describes these dual frames of frontier making as 'bountiful emptiness' (Bridge, 2001). A frontier is depicted as full of potentials and opportunities while imagined as unmapped and unpopulated. Such dialectic formation of frontiers has an effect of legitimizing interventions and exploitations. Yet, interventions often fail. As Jason Cons and Michael Eilenberg note in the Introduction of this volume, frontiers often become 'terrains of failure, contestation, and conflict'. On China's coasts, 'imposing sustainability' creates, if not outright failures, tensions, frictions, and anxiety that reveal the unsustainability of the very interventions carried out as sustainable (Neumann, 1998).

Focusing on Tianjin Binhai New Area (TBNA) and Tangshan Caofeidian, two adjacent large-scale reclamation sites in the Bohai Bay region, I interrogate the controversies around this particular vision of sustainability that have been projected onto the coasts. While I draw upon a literature on China's eco-cities that has critiqued China's embrace of global discourses of sustainability, my observations reveal that the processes of spatializing sustainability reach beyond eco-city sites and spread along the entire coast. I rely on an assemblage approach to develop a critical analysis of sustainability. In doing so, I highlight the paradox of China's sustainability agenda which allows interventions to erase existing biophysical and social realities on the coast while writing new ecological futures upon the subsequent coastal ruins (Stoler, 2013; see also Paprocki, this volume). Navigating through reclamation landscapes raises fundamental questions about 'sustainability' of what, and for whom. Addressing such questions, this chapter makes a case that China's coasts have turned into a *contested* frontier of sustainability.

China's Coasts as a Frontier Assemblage

The nationwide reclamation fever that sprouted in the early 2000s had gone almost unnoticed by the public for many years. It was only after the Tangshan government's debt scandal in 2013 regarding its inability to finance the Caofeidian reclamation project that major Chinese media outlets began to pay attention to large-scale reclamation practices in the country.[4] A year later news spread globally with a Guardian report that delivered vivid visual presentations of Caofeidian that by that time had turned into a 'ghost city' (see Woodworth, this volume).[5] That China's coasts had not pre-existed as a major source of land and that coastal reclamation practices sporadically but contemporarily became a prominent way to provision urban and industrial land suggest what Bruce Braun called 'the contingency of the present' (Braun, 2002). An assemblage, according to Braun, is a transitory moment where things do not pre-exist but create a contingently and temporarily fixed state. China's coasts have emerged through a frontier assemblage of heterogeneous events, processes, and forces converging against the backdrop of China's developmental and environmental challenges. Among the multiple elements that constitute this assemblage, I here elaborate two at work at the national level: a desire for land and a desire for spatializing sustainability.

A desire for land

Land is a peculiar commodity. Land is geographically fixed and as such is unable to be removed. As Tania Li describes it, 'you cannot roll it up and take it away' (Li, 2014b: 589). Likewise, land does not reproduce.

Land may be repackaged for the purpose of more intense and diverse uses, but the sources for new land are strictly limited. Land in China has an extraordinary status. Despite the fact that China has the world's third largest land area as its territory, land is considered one of the nation's most critically *scarce* resources. In popular discourses, land scarcity is linked to population increase in markedly Malthusian terms: through fears of shortages in food and housing provision. For example, an often-invoked narrative says that the size of arable land per capita in China is merely 25–41% of the global average.[6] Land scarcity is narrativized to justify interventions to protect agricultural land. Similarly, a need for housing to accommodate the growing population justifies the desire to seek additional land.

These two types of land use (agriculture and housing) have been in intense conflict over the past decades because of China's rapid urbanization. During what You-Tien Hsing termed 'the great urban transformation' (Hsing, 2010), the urban population rocketed – from 20% in 1978 to more than 50% in 2012. The figure is expected to rise to 70% by 2030. In terms of physical space, land designated as 'urban' expanded dramatically, at a rate even higher than the growth of urban population (Development Research Center of the State Council and World Bank, 2014). As land was mostly provisioned through rural-to-urban land conversion, rampant and forceful conversion of agricultural land led to massive displacement of rural residents. A tense period of massive rural demonstrations against urbanization in 1997 even forced the government to declare a one-year moratorium on land use conversion (Cartier, 2001; O'Brien, 2008).

The coastal land-reclamation boom of the 2000s is primarily situated in this struggle between urban and agricultural land. It is uncertain whether the new regulations on farmland protection were driven by the government's political fear of potential insurgencies by dissatisfied farmers or by its concern about the decline in agricultural production due to unregulated land grabbing. Nevertheless, the late 1990s and the early 2000s saw a rise of discourses invoking potential food insecurity, which effectively legitimized the government's hurried move and strict enforcement of agricultural protection. The New Land Administration Law, set up in 1999, designated a total of 1.2 million km^2 of agricultural land as protected basic farmland across China (Lin and Yi, 2011; Lai et al., 2014). The outcome was the creation of what is called the 'red-line': a firm boundary where urban growth must stop before encroaching onto rural land.

The notion of land scarcity translated into stricter farmland protection policies hence set up the stage for prompting coastal land reclamation as an alternative source of land. This in turn became a major source of income for local governments. A study reports that up to 70% of local governments' revenues were estimated to come from land development in 2010; the figure is around 45% as of 2014.[7] With this imperative for profit-making from land development, local governments frequently invoke the term 'red-line' when asked about their rationales for delving into large-scale reclamation. 'If you want to move the red-line, you must compensate hundreds of thousands of yuan for the cost of farmland', I was told in 2013 by a prefecture-level official in charge of coastal reclamation.

But for what do they keep producing land? Today's coastal reclamation fever is distinguished from previous ones not only for its scale but for land use. Reclamation booms prior to the 1990s were driven by the production of fish farms, salt paddies, and agricultural land. In contrast, official statistics indicate that over 90% of the new land reclaimed since 2002 is earmarked for 'construction land', i.e. for urban and industrial purposes.[8] Given the scale of post-millennial reclamation practices, this indicates that the desire for land is specifically about what Myung-Rae Cho has called 'construction as accumulation' and, in particular, the spatial expansion of urban development (Cho, 2006). In this sense, discourses of land scarcity that drive reclamation are *relative* ones: the demand for urban land from reclaimed coastal spaces was a reaction to the heightened institutional barrier to accessing agricultural land, while the underlying cause of urbanization was never questioned.

A desire for spatializing sustainability

Circa the late-2000s, a new phenomenon began to emerge where coastal land reclamation was increasingly labelled as an 'ecological' practice. In other words, reclamation projects were increasingly reimagined as sound 'eco' projects that were good not only for growth, but also for the environment. 'Eco-' labels flourish in reclamation sites today. It is not cities, but 'eco-cities' that are built on reclaimed land. New industrial zones are justified by their reduction of greenhouse gas emissions, waste, and pollution and their increased recycling functions for realizing a circular economy (Zhang et al., 2011). Renewable energy infrastructure as well as the engineered nature in the form of canals, wetlands, and green landscaping have become integral to coastal land-reclamation projects.

This 'environmentalization' or 'greening' of coastal land reclamation is a key feature of the state's broader sustainability agenda (Chen, 2013; Anderson, this volume). This agenda seeks to revamp the very configuration of society in order to secure 'harmonious, balanced, and sustainable development'. The agenda can be traced back to 'eco-civilization', a slogan announced by Hu Jintao at the 17th National Congress of the Communist Party in 2007. Julie Sze notes that Hu's speech was 'the first time the Chinese Communist Party had highlighted ecology, putting it explicitly on the Party's agenda' (Sze, 2015: 36). Since then, the Chinese state has pursued this goal aggressively. China's past two national five-year plans adopted various mechanisms including green finance, an emission trading scheme, and investment in expanding renewable energy infrastructure.[9] While the question remains as to how many of these mechanisms will be effectively enforced, an increasing number of commentators agree with Jane Golley's claim that 'there is ample evidence to suggest this commitment is real' (Golley, 2016: 7). Golley argues that China's capacity to bring changes has reached 'an extent that green supporters in advanced democratic countries can only dream about' (2016: 7). China's push for sustainability is now envied and celebrated in various venues, as demonstrated by a recent World Economic Forum video boasting that 'China is now the world's biggest producer of solar power'.[10]

'Sustainability' as a malleable, undemonstrated concept crucial to the environmentalization of the Chinese state necessitates space to test its possibilities and to prove its feasibility and efficacy (Chen et al., 2017). In China, new eco-cities have assumed that role. Eco-cities are a globally emerging urban planning model designed to save energy and minimize environmental impacts. Eco-cities have been enthusiastically embraced by the state as a solution to the multi-faceted socioecological challenges emerging from decades of industrialization-driven development (Caprotti, 2016). Moreover, eco-cities allow the Chinese state to continue to pursue sustainability without giving up an urbanization-driven growth model. In effect, eco-cities are at the centre of the national land development strategy. As Chien argues, eco-cities are 'the third round of large-scale transformation since the economic reform' following the development zones (kaifaqu) in the 1980s and college towns (daxuecheng) in the 1990s (Chien, 2013). These indicate that eco-cities are themselves an assemblage formed at a historical conjuncture of the material consequences of intense industrialization of the past, the ongoing urbanization-driven growth strategy, and China's emergent vision for a sustainable future.

Many authors have studied eco-cities constructed on reclaimed land, although they have not focused on processes of reclamation per se (Chang and Sheppard, 2013; Joss, 2015; Sze, 2015; Caprotti, 2016). Yet the two processes are intricately entangled in ways that are beyond the political economy of land that renders coasts an attractive source of land. The new state agenda of sustainability is less tolerant about the massive scale of ecological destruction that inevitably accompanies reclamation practices. Destruction needs to be justified by producing something countervailingly desirable. Eco-cities fulfil such a gap. In other words, the making of a sustainability frontier is *essential* for legitimizing the desire for land. This way, the desire for spatializing the vision of sustainability and the desire for land intersect through reclamation on China's coasts.

A Contested Sustainability Frontier

The Binhai New Area in the Tianjin municipality and Caofeidian in the prefecture-level Tangshan City in Hebei are among the largest coastal reclamation sites in China. The story of the two reclamation sites, despite their differential administrative status and reputation, illustrates several characteristics shared by numerous other sites along China's coasts. Specifically, they both feature large-scale land development driven by the strong will of local governments, an often-prolonged construction period due to financial constraints, and mixed industrial and urban planning. Both include an eco-city plan which effectively 'greenwashes' the reclamation project site.

In the TBNA and Tangshan Caofeidian, local governments' political motivations grounded on the notion of 'lagging behind' prompted large-scale reclamation. At the regional level, there was a widespread perception that the Jingjinji (Beijing–Tianjin–Hebei) metropolitan region had been losing competition compared to the dazzling success of the Pearl River Delta region. At the city level, the rapid growth of other provincial-level municipalities such as Beijing and Shanghai in the post-reform era prompted the public sentiment that 'Tianjin became an underdeveloped city, lagging behind Beijing, her neighbor, during China's 30 years of planned economy' (Zhu and Sun, 2009: 195). Similarly, the Caofeidian project allegedly was propelled by the Tangshan government's jealousy of Tianjin's relative success.

In this context, reclamation was part of broader development planning for boosting economic growth in both sites. The TBNA is composed of old and new development areas, including three major industrial zones (Tianjin Economic-Technological Development Area, Tianjin Port Free Trade Zone, and Tianjin Port), three administrative districts (Tanggu, Hangu, and Dagang), and the Sino-Singapore Tianjin Eco-City. About 307 km² of coast was reclaimed as new development over the past decade with more planned in the near future (Zhu et al., 2017). Tangshan, on the other hand, envisioned its own development planning after it lost to Tianjin in the national bid for the siting of the Sino-Singapore Eco-City. The city government largely relies on coastal reclamation for land provision and on bank loans. The Caofeidian (CFD), like the TBNA, is a comprehensive land planning project including the ambitious Caofeidian International Eco-City (150 km²). About 310 km² of reclamation is planned; about 230 km² has turned into land so far (Wang et al., 2014; Liang et al., 2015).

But perhaps the key dynamic in assembling the coast as a frontier is the civil engineering process of making land. As recent infrastructure studies demonstrate, large-scale civil engineering practices not only bring transformative socioecological change but do so by enhancing irreversibility in such a change (Graham, 2010; Carse, 2014; Harvey and Knox, 2015). Land reclamation is a multi-stage process – necessitating building a seawall, drying the ground inside the wall, filling in the ground with such materials as sand, mud, and construction waste, and pounding and desalinating the elevated ground. The process of reclamation creates messy and dusty landscapes and irreversibly buries marine life and habitats underground. For these reasons, elsewhere in Asia, particularly in South Korea, coastal land-reclamation projects were intensely opposed by environmentalists for this brutal materiality of land-making (Hahm and Kang, 2007; Park, 2007; Choi, 2014).

In the TBNA and Caofeidian, as is in many other sites in China, significant civil opposition has not occurred for a variety of reasons. Yet, the dual tasks of frontier making – conjuring the dream of a sustainable future and rendering the previous environment as undesirable – are still ongoing. Apart from the various venues such as websites and government documents, visitor centres seek to make the imagined future a tangible reality. The visitor centres of the Tianjin Sino-Singapore Eco-City and the Caofeidian International Eco-City embody the desire for sustainability with overwhelming representations of 'green' and 'ecology'. For example, the Tianjin Eco-City visitor centre welcomes visitors with a picture of egrets and lush wetlands covering an entire wall of the main hall. The miniature planning model of the TBNA, not just the eco-city, is full of trees and waterways. In contrast, the past landscape is narrated as a polluted wasteland. For example, a series of images tells the tale of transforming a heavy-metal-contaminated pond into a clean waterfront. The visitor centres deliver a carefully purposed message that the grey construction landscape outside the building is merely transitory and that a new blue and green future is will soon become a reality. In doing so, they hide the destructive dimension of reclamation; instead, reclamation is presented as a desirable practice that rescues nature from the existing status of heavy pollution and even lets it prosper. In this way, the visitor centres display the transformation of China's sustainability dream into reality. In advancing this agenda, they encourage the visitor to see past the actual scenes they see and instead envision a green and sustainable future (Figure 7.1).

Figure 7.1 Tianjin Sino-Singapore Eco-City visitor centre. Photo: Young Rae Choi.

Increasingly, efforts to create a sustainability frontier on reclamation sites requires a rewriting and memorialization of local histories, fishing traditions, and cultures. This produces an ironic situation in which the fisher communities and marine ecologies that have been destroyed through reclamation are remembered as those in need of preservation. For planners, they are locally-specific assets for tourism, which is considered as a less-polluting and profitable 'eco-industry' squarely fit for a sustainable future. For example, within the TBNA, the Tianjin Tourism Area next to the Tianjin Sino-Singapore Eco-City enshrines a 42.3 m-tall statue of Mazu, a sea goddess known as a patron of fishers and sailors. The statue was erected in memory of the very culture and tradition of fishing destroyed through reclamation (Figure 7.2).

The gigantic sea goddess powerlessly watching over waters being transformed into dry land signifies a troubling vision of sustainability. The frontier of sustainability, specifically construed as an urban, energy-efficient, and recreational utopia for some, is a dystopia for others. While a majority of fishers moved away from the two reclamation sites by the time of my research, a few of them remain in Zuidong, an ad hoc development project site of the greater Caofeidian area. They attest to how limited the government's policy for the displaced is. Seasonal fishers whose hukous (household registers) are elsewhere, regardless of how

Figure 7.2   Statue of Mazu, a sea goddess known as the protector of fishers and sailors, in the Tianjin Tourism Area. Photo: Young Rae Choi.

long they have been engaged in fishing in the area, received no compensation for their displacement. The residents of Zuidong on the other hand were forced to relocate to new housing provided by the city government, which already had developed architectural defects. Relocation to a town and acquisition of an urban hukou, which generally is understood as a more attractive option than remaining in a rural village, can present substantial challenges to fishers who are not accustomed to the land-based culture of regular incomes and expenditures. An elderly shrimp-oil seller in Zuidong, for example, worried that he would not be able to pay monthly rents and utility fees for the relocation housing.

Raising the question of whether the new spaces of sustainability along China's coasts will actually achieve a sustainable future contests the very concept of 'sustainability'. First of all, there is an impending resource crisis. An urban planner at the China Academy of Urban Planning and Design who participated in the planning process of Caofeidian told me that the large green spaces built into the eco-city design are worsening chronic groundwater shortage problems in the region. The Tianjin-Tangshan region is already running two desalination plants with a plan to build another.[11] Meanwhile, an employee of the Tianjin Sino-Singapore Eco-City told me that that renewable energy-generating streetlights, which were built even before the construction of buildings that will use the electricity, have sat idle for a prolonged period of time as the reclamation projects were delayed. In the meantime, for maintenance against corrosion, they require the use of extra electricity.

The vision of sustainability embodied in land reclamation also saddles local governments with massive financial burdens. When the anticipated financial flow does not materialize – when reclaimed land remains unsold – the material realities of sustainability could helplessly shatter. In May 2013, Caofeidian received nationwide attention as a news reporter disclosed the financial plight of the project. The report estimated that the Tangshan government's debt would be over 60 billion RMB (equivalent to US\$9.25 billion), as promised private investment was cut in half in just two years from 2009 to 2011.[12] It was not only Caofeidian that suffered from the slowdown in the Chinese economy after the 2007–2008 global financial crisis. But the project's financial hardship stood out in the absence of higher-level political commitment in comparison to the TBNA. When I visited the site several months after the news, the Caofeidian International Eco-City construction had already stopped.[13] A spacious upper-middle class housing complex designed to mimic an American suburban neighbourhood was entirely abandoned. The roofs were broken and the mini solar-wind energy-generating stands were corroded. Weeds had grown high. At the outward edge of the eco-city, intended to be another luxurious waterfront residential area, teals had hatched eggs on dusty, desert-like land created with mud pumped from the sea bottom just outside the seawall. According to a 2014 survey conducted by the Let Birds Fly Public Fund, over 40% of businesses and industries in Caofeidian were also either closed or left uncompleted (Figure 7.3).[14]

Taking the examples of the Caofeidian International Eco-City and another never-built Dongtan Eco-city in a suburb of Shanghai, Changjie Zhan and Martin de Jong argue that financial failures of eco-city projects in China are 'not uncommon' (Zhan and de Jong, 2017). Studying the case of the Sino-Singapore Tianjin Eco-City which they claim to be 'the best-known and arguably the most successful large-scale sustainable newtown development project in China', the authors try to find solutions to secure financial sustainability of the supposedly sustainable cities (2017: 201). Yet, even the Sino-Singapore Tianjin Eco-City, let alone the entire TBNA, remains only partially complete and years behind the initial planning schedule despite the fact that 'the stakes the Chinese national and Singaporean governments put in it are so high that they will do everything to make it a success' (2017: 213). The juxtaposition of Caofeidian and the TBNA show that the coastal sustainability frontier sustained by the dream of profit is inescapably a contested space.

Figure 7.3 Abandoned houses, Caofeidian International Eco-City. Photo: Young Rae Choi.

Conclusion

In this chapter, I have explored how China's coasts have become an essential part of the Chinese state's sustainability agenda. Moreover, I have examined how reclamation sites that sit at the crossroad of the desire for land and the desire for sustainability become troubled spaces. The transformations of the coastal landscapes and the reconfigurations of coast-society relations in twenty-first century China are not isolated events. They are the outcome of complex and contingent forces in Chinese political economy. The complexities produce the conditions to project and materialize a vision of sustainable futures onto the present. These new visions not only mask the destructive outcomes of coastal land reclamation, reconfiguring them as a productive ecological practice. They also create particular and troubling new landscapes and materialities.

The concept of frontier assemblage helps interrogate and expose meaning making within and around coastal space. As I have shown, new debates around land-reclamation produces China's coasts as a sustainability frontier. They do so, first by 'unmapping' spaces laden with rich human and non-human histories (Tsing, 2003; Roy, 2011; Lentz, this volume); and second by remapping an idealized 'dream of eco-development' onto the spaces that are made empty through destructive reclamation (Sze, 2015). In the process, discourses of sustainability re-write local histories for the promotion of ecological tourism.

The story of Mazu, the sea goddess cast as a statue in the Tianjin Tourism Area, does not end there. Originally from Meizhou, a small island in Fujian province, Mazu is most popular in Taiwan and the southern provinces of China.[15] In contrast, she is not a commonly revered figure in the Bohai Bay. Her temples in Northern China are mostly small and modest, in stark contrast to the lavish and colourful ones in the South China Sea where she is dearly worshiped. Nevertheless, here she is, enshrined in a park named after her, instead of a temple, by the TBNA tourism department. The Mazu Cultural Park is one of the tourism development projects associated with the efforts to create a popular Binhai brand.[16] This foreign, out-of-place goddess is painfully looking over the fishers in predicament, who, it turns out, are not her own people in the first place. The place where the remaining fishers would be driven out eventually will be filled with tourists who come to appreciate the lost cultures of the seas. This is a deeper irony of what is meant by a sustainability frontier.

Notes

1   See Declaration of International Workshop on Intertidal Wetland Conservation and Management in the Yellow Sea Provinces of China. (2014). In: *International Workshop on Intertidal Wetland Conservation and Manageent in the Yellow Sea Provinces of China*, 2. Beijing.
2   State Oceanic Administration. (2003–2013). Sea-Use Management Report. www.soa.gov.cn/zwgk/hygb/hysyglgb/201403/t20140320_31036.html.
3   For example, see UNDP/GEF Yellow Sea Project (2007) and MacKinnon et al. (2012).
4   See Liu, Y., *et al.* (2013). Tangshan Caofeidian construction suspension warning. *Twenty-First Century Economy Report* (25 May). http://finance.qq.com/a/20130525/001268.htm.
5   See Sabrie, G. (2014). Caofeidian, the Chinese eco-city that became a ghost town – in pictures. *The Guardian* 23 July:1–16. https://www.theguardian.com/cities/gallery/2014/jul/23/caofeidian-chinese-eco-city-ghost-town-in-pictures.
6   See Ministry of Land and Resources of the PRC. (2004). *Communiqué on land and resources of China 2003*. www.mlr.gov.cn/mlrenglish/communique. See also Smil (1999) and Chen (2007).
7   See Development Research Center of the State Council, and World Bank. (2014). *Urban China: Toward efficient, inclusive, and sustainable urbanization*. World Bank Publications.

8   See State Oceanic Administration 2003–2013.

9   CCICED. (2016). *Interim Report of 'China Green Transition Outlook 2020–2050' Project.*

10   World Economic Forum. (2017). China is now the world's biggest producer of solar power. https://www.youtube.com/watch?v=YB63GYAXe50.

11   Wong, E. (2014). Desalination Plant Said to Be Planned for Thirsty Beijing. *The New York Times* (15 April). http://www.nytimes.com/2014/04/16/world/asia/desalination-plant-beijing-china.html?_r=0.

12   See Liu, Y., *et al.* (2013).

13   See Sabrie (2014); Vanderklippe, N. and E. Reguly. (2014). China's looming debt bomb: Shadow banking and the threat to growth. *The Globe and Mail* (3 May). http://www.theglobeandmail.com/report-on-business/chinas-looming-debt-bomb-shadow-banking-and-the-threat-to-growth/article18409275/?page=all.

14   Zhou, H., S. Chen, D. Huang, W. Jin, and Y. Wang. (2014). *Intertidal zone loss, migrant birds have nowhere to rely on – Yellow and Bohai Sea coastal reclamation sites survey report.*

15   Wu, M., and Z. Li. (2017). Tianjin Binhai New Area: Rising Manufacturing Center and Tourism Destination. *China Today* (2 March). www.chinatoday.com.cn/english/culture/2017-03/02/content_736432.htm.

16   Wang, N. (2017). Tourism development in TBNA prioritizes 'Binhai Brand'. *Tianjin Binhai New Area Government.* http://english.tjbh.com/system/2017/08/14/030259979.shtml.

8

Spaces of the Gigantic

Extraction and Urbanization on China's Energy Frontier

Max D. Woodworth

Introduction

This chapter seeks to understand energy resource exploitation and urbanization as key forces driving an unrelenting process of transformation in China's Inner Mongolia Autonomous Region in the 2000s. From the enormous mining operations dotting its rural areas, to the expanding networks of pipelines and transport infrastructure, to the massive urban projects arising throughout the region, Inner Mongolia's sudden ascendance as a socially produced keystone space in China's energy supply is vividly inscribed in its physical landscapes, shaping it into what I call a space of the gigantic. These changes evince the impacts of differently scaled and intersecting spatial planning regimes and policy shifts, and they show the capacity of these to reshape physical and social environments in the name of 'development'. At the same time, Inner Mongolia's radical transformations have served to dramatically re-script this vast region from a space of marginality and under-development to one of spectacular, yet precarious, centrality and economic dynamism, as well as ecological crisis (see also Choi, this volume).

Following other contributors to this volume, this study is situated within the broad concerns of the political economy of Asia's frontier transformations, particularly the relationships among space, resources, and different forms of social power. In recent years, scholars in various fields and geographical contexts have also revived the frontier concept to interpret a range of developmental processes at the intersection of

Frontier Assemblages: The Emergent Politics of Resource Frontiers in Asia,
First Edition. Edited by Jason Cons and Michael Eilenberg.

these diverse themes (Moore, 2000; De Angelis, 2004; Barney, 2009; Watts, 2014, 2015a). Within this literature, the frontier represents a peculiar space of temporally compressed, spatially convulsive, and conflict-ridden change. Such studies underscore the importance of attending to the socially constructed peripheral-ness of frontiers as effects of what Doreen Massey (1994) called uneven 'power geometries' within national and global economic structures. In short, frontiers are not pre-existing regions at national or civilizational margins (as in the classical Turnerian 'frontier thesis') but rather are spaces actively produced by dynamic combinations of actors and institutions interacting and colliding in the making of restless spaces whose utility is precisely their paradoxical combinations of connection and disconnection (see Günel; Anderson; and Zee, this volume). As the editors of this collection note, this sense of the frontier as a conjuncture of forces permits a view of such spaces as more than merely awaiting incorporation; they are also creative sites where new social forms emerge through unstable combinations of bottom-up action and top-down coercion. It is in this sense that this chapter understands the notion of 'frontier assemblage', as emergent and social-spatially contingent convergences of forces and people that result in specific, yet never predetermined outcomes (see Ong and Collier, 2005; Li, 2007b). Such a notion of the frontier assemblage is germane to a theoretically robust understanding of the politics of invigorated extraction and city building that are reshaping Inner Mongolia along with much of the rest of western China.

In this chapter, I am especially keen to evaluate a consistent but neglected feature of frontier landscapes, namely the hyperbolic – or, 'gigantic' – qualities that characterize much of the thinking behind, as well as the concrete outcomes of, development efforts in such sites. As the critical frontier scholarship just mentioned has convincingly shown, frontier formation is premised upon complex social-spatial relations favouring what Jason Moore calls 'hyper-exploitation' (2000). In a series of essays on the topic of the frontier, Michael Watts has argued that the formation of the frontier 'should not imply only the technical relations of resource exploitation (as the industry understands frontiers) ... Rather', he continues, the frontier 'marks the construction of a new space of accumulation and creation of the conditions of existence for' systems of massive-scale extraction (2014: 194). This entails the refinement of the technical and institutional methods of production as well as the creation of social conditions ripe for exploitation. Moreover, Watts avers, a view of the frontier is incomplete without also accounting for the ways in which it exists representationally, or as a figment of diverse discursive components. Exploring the oil frontier, he remarks 'oil is not simply part of but is represented, experienced, and partially constituted by an archive of cultural productions and especially by a vast, often spectacularized image world' (2015a: 167). In what follows, I show how making the conditions of possibility for hyper-exploitation eventually comes to permeate the regional environment beyond the spaces of extraction, becoming part of how the frontier space is organized, remade, and broadly understood. More concretely, I aim to show how designs for massive-scale mining in the region connect with urban megaprojects, speculative fevers, and new imaginative frames that seek to represent frontier change. The broader argument of the chapter is that the frontier must be understood as more than just a recipient space, a zone impacted by inward investments and violent enclosures, but also as a space that generates material and social transformations locally that are in dialogue with the forces that drive hyper-exploitation. As such, the frontier is a distinctive and uniquely vital space because of its confounding mélange of forceful dominance and unregulated free-for-all, a combination that seems to obviate constraints on the scale of exploitation and that inspires, by turns, enthrallment and feelings of intense catastrophism (see also Paprocki, this volume).

The Frontier as Space of the Gigantic

In recent years, a number of studies have sought to analyse the political and cultural connotations of massive landscape changes in China. Pointing to megaprojects and post-modern architectural landmarks in Beijing or Shanghai, Aihwa Ong heralds a 'new cultural regime' (2011: 205) of urban spectacle that advances geopolitical goals of authoritarian centralized states. It is worth pausing, however, to consider how other spaces conform or diverge from this ostensible regime. Li Zhang's (2006) study of Kunming, for example, has shown how deeply felt anxieties about temporal and spatial 'lateness' spur local governments in lower-tier cities to favour massive-scale urban reconfigurations as part of explicit catch-up modernizing agendas. In Xinjiang and Tibet, as well, research has shown how new landscape change is geared less toward proclaiming 'world city' status than broadcasting the Chinese state's colonial claims to contested territorial fringes (Cliff, 2013; Yeh, 2013). Recent work by Franck Billé (2014) and Tim Simpson (2008) has also explored China's margins as spaces onto which a multiplicity of agents project myriad conceptions, dreams, and myths of (national) development and modernization.

At stake in Inner Mongolia are the specific ways that spatial transformations differ in reflection of the region's status as a new extractive frontier. This is particularly urgent as central-government programs, notably the Opening the West (*xibu da kaifa*) campaign, have sought to restructure national space into a bifurcated geography of resource production and consumption with Inner Mongolia figuring centrally as a vital supply region. The very names of programs advanced under the banner of the Open the West program highlight its geographical basis: 'south-north water diversion' (*nanshui beidiao*), 'west-east gas delivery' (*xiqi dongsong*), 'west-east oil delivery' (*xiyou dongsong*), 'west-east electricity transmission' (*xidian dongsong*), 'north-south coal delivery' (*beimei nandiao*).[1] Each of these discloses a spatial division of labour in which certain places are targeted for intensive extraction. Indeed, as a constellation of extractive activity, these development efforts plot the contours of 'geological provinces' (Watts, 2014) and 'powersheds' (Magee, 2006) whose functions are to provision the urban-industrial expansion of the east with a range of vital natural resources. On the ground, this regional transformation takes the form of massive surface mines, pipelines, canals, dams, anti-desertification barriers, and, not coincidentally, spectacular urban megaprojects in an archipelago of resource boomtowns that stretches the length of Inner Mongolia. Local landscapes thus reveal the confluence of regional development agendas as well as eminently local visions of urban transformation animated by a pulsing excitement exhibited by policy makers, local-government leaders, and a significant share of local populations toward rapid, resource-led growth.

Fundamentally, China's regional resource-supply agendas reveal an oculo-centric view of space, one that seeks, in James Scott's terms, to miniaturize and render legible what are, in fact, socially dense and physically vast areas (1998). The requisite imaginings and physical remakings of landscapes under these conditions echo Susan Stewart's (1993) incisive reflections on the ideological content of gigantic and miniature forms. The gigantic, she argues, is the quintessential mode by which concentrated political power achieves symbolic form in the physical environment, thus making it both a display and exercise of power (1993: 81). The gigantic manages this dual function by enveloping and miniaturizing the body while also remaining inaccessible. Because it fills the space beyond the body, Stewart argues, the gigantic 'analogously mirrors the abstractions of institutions – either those of religion, the state, or, as is increasingly the case, the abstractions of technology and corporate power' (Stewart, 1993: 102). In these ways, it minimizes the individual by reducing her to the position of passive spectator. The gigantic thus symbolizes a power relation mediated through huge objects and spaces, both of which the frontier features in super-abundance. But to suggest that something is obviously 'big' or, as it were, 'gigantic', is to make a claim about relations in space. For as John Law (2004) notes, scale and size mean little outside their social context. 'If [something] is bigger or smaller then it is because it can be made bigger or smaller in this site or that' (Law, 1994: 13; quoted in Jacobs, 2006: 14). Hence, the materiality of the large is inseparable from the manifold ways it is filtered through spatial practices and culture, thus making it a structure of power and 'a way of seeing'.

Cultural geographers have charted a view of landscape in similar terms in recent decades (Cosgrove and Daniels, 1988; Mitchell, 2008). Landscapes are always ideological; and the landscape, especially that of the city but also of the resource frontier, is a prime example of the gigantic in action. As Ey and Sherval (2016) note, the 'minescape' is a rich discursive as well as material repository. In these ways, Stewart's notion of the gigantic reverberates the production of frontier spaces, where massive size becomes a persistent way of imagining and delivering spatial change, playing on and reifying the conceit of abundant empty spaces and intimating the allure of limitless and unobstructed growth. With this in mind, the gigantic might be seen, on the one hand, to reflect political impulses toward control of resource-abundant borderlands, while, on the other, paradoxically throwing fuel on the fires of specula-tive bonanzas. At the same time, however, the gigantic also helps gen-erate an aesthetic in physical space that comes to symbolize ruin as readily as it does prosperity and that elicits a diversity of responses at different moments across boom-bust cycles.

Returning to the concept of the frontier, Anna Tsing has described such spaces in the following terms: 'Frontiers are not just discovered at the edge; they are projects in making geographical and temporal experi-ences' (2003: 5100). It is these experiences in reshaping and re-casting spaces as frontiers, and the ways supreme size figures so centrally in these changes, that I explore in the subsequent sections. I report on them in the form of dispatches from fieldwork across Inner Mongolia coupled with data drawn from the archive of exploitation that the region's recent transformations have produced.

Unpacking the Frontier

A dispatch from the pits: Upon first entering the Heidaigou surface coalmine, which, along with the nearby Ha'erwusu mine in the Jungar region of Inner Mongolia, forms the largest surface coal-mining complex in the world, I experienced a curious sense of role reversal. What I felt was a palpable sense of becoming toy-like, as though reduced to a figure in a giant's sandbox. Having joined a machinery repair crew in the mine during fieldwork, the men in the repair team recognized my initial *disorientation and reassured me that I would eventually become accustomed to the sheer size of the place. The most important thing, they warned, was to follow safety protocol, because in the mine, people are so tiny that they can become essentially invisible to machine operators. In surface mines, accidents tend to occur as encounters with rolling and moving machines, rather than the traditional hazards of roof collapse, floods, and methane build-up that make underground coal mining still far more lethal. The mine's service roads that afford entry into the mine are up to two hundred meters wide in places in order to safely accommodate the enormous dump trucks that ply them day and night, but also to allow passage of the even larger machineries – draglines and bucketwheel excavators – that are the true workhorses of the mine. An elaborate choreography of massive machines in the mine is coordinated by unique traffic rules, movement patterns, and staccato warning blasts of truck horns. To minimize dust and to ensure safety, everything in the mine moves slowly – or at least appears to move slowly given the sheer scale of the vehicles. The dump trucks, service vehicles, drill rigs, and buses that ferry workers to various spots throughout the mine are unmistakably subordinate to these mega-machines that are the only things that appear proportionate to the space. Excavating machinery dwarfs even the 150-ton load capacity dump trucks, which in turn tower above our extended-cab pickup truck. At the bottom of the pit, some two hundred meters below the surface, I take moments during the day to approach and touch the coalface, which forms a stratum of shiny black bituminous rock looming 25 m thick overhead and extending in a row hundreds of meters long. This is the active coalface, where the excavation of the coal is currently underway. Eventually, more of the overburden will need to be removed to get at the rest of the coal seam, which spreads out across an area of 67 km^2. In a former time, likely around 300 million years ago, this stratum was a steamy bog near the equator. Now, through coordinated dynamite blasts that open the Earth's surface like a zipper, and with the loosened material moved aside by gargantuan machines, the stratum is exposed once more, its contents loaded onto conveyor belts and whisked up and out of the mine on its way to power plants. My direct contact with this material index of geological time is cut short by the teeth of digging machines, as the rock is carted away for combustion.*

Critical examinations of frontiers face the challenge of specifying where, ultimately, they are located. Like the analogous term 'region', the frontier has an elusive geography, a slipperiness that is at the heart of understandings of them in relational terms. As Keith Barney (2009) has argued, the frontier must be regarded as a relational space, meaning one that is composed of connections and interactions across territories and times. As such, they are 'dynamic and unstable', a 'permanent prospect' (Watts, 2014: 193) defined by perpetual disappearance and emergence (and, often enough, re-emergence). And yet the resource frontier does have some locational coordinates at points in time. In this sense, the Heidaigou coalface is the ground zero of China's energy frontier in the current moment, an intersection point of geological and economic temporalities and of physical and social spaces, whose collisions here have rendered the sub-surface materials into vitally important sources of profit and power, both electrical and social.

The origins of this eventuality might be traced back millennia, but it suffices to note its more recent lineage. In a 1956 speech on the 'Ten Great Relations', Mao Zedong noted: 'The population of the minority nationalities in our country is small, but the area they inhabit is large ... We say China is a country vast in territory, rich in resources and large in population; as a matter of fact, it is the Han nationality whose population is large and the minority nationalities whose territory is vast and whose resources are rich, or at least in all probability their resources under the soil are rich'.[2] Though Mao's full remarks addressed the politics of inter-ethnic relations in the newly established People's Republic of China, where energy resources are concerned, his intuition about the rich resource endowments of western China was on the mark. In the key category of coal, the leading energy source by far, China has the world's third-largest proven reserves at 186 billion tons (GT). Yet these resources are heavily concentrated in the poor and isolated parts of the west: 67% are in Shanxi, Shaanxi, Inner Mongolia, and Ningxia (Huang, 2010; Tu, 2011; Li, 2013). Inner Mongolia alone has an estimated 800 GT of coal in fields located throughout its vast territory, suggesting the province could radically boost China's recoverable reserves with the commensurate advances in production technologies. The autonomous region also boasts a third of China's natural gas reserves, a portion of China's scant domestic oil deposits, and theoretically massive wind and solar power potential that is now beginning to be exploited.

Geological conditions such as these have underpinned aggressive resource exploitation strategies for Inner Mongolia dating back decades and leading to the inauguration of the Open the West campaign in 2000. From the outset, the Open the West program has sought to take advantage of the combination of relative regional poverty and under-development in the west and turn it into an advantage through large-scale resource exploitation. In Inner Mongolia, the Open the West campaign called for accelerating and increasing the scope of coal mine projects, expanding exploration and production of oil and gas, and dramatically expanding the network of pipelines to deliver these fuels to markets east and south. Guiding principles, according to the campaign's key architect, were to increase primary energy export capacity and intensify resource processing in extraction regions to enhance the value added to primary resources (Zeng, 2010: 115–117). Topographical conditions favourable to surface mining in Jungar but also in Xilin Gol, Wuhai, and the Dongsheng field in Ordos Municipality further made expansion in Inner Mongolia a priority (Huang, 2010).

As a practical matter, however, the energy resources encased in the soils of Inner Mongolia had remained largely out of reach until recently (Tu, 2011). This was due to the prohibitive cost of moving heavy, low-value primary goods like coal across the large distances separating reserves from major markets located predominantly in the east and south of the country. At prevailing prices through the 1990s, energy resource production in Inner Mongolia was inconsistently profitable and remained marginal to the national supply scenario. Also critical here was the under-development of pipeline and freight rail, which stunted the development of the local industry. The limited capacity of China's electrical grid further militated against *in situ* power generation in the territorial peripheries. As a consequence of these factors, Inner Mongolia's coal production was characterized until recently by fragmentation and under-capitalization, low mechanization rates, scant oversight, severe environmental degradation, high casualty rates, and serious waste of resources (Huang, 2010). From the perspective of central planners in Beijing, these conditions represented problems in need of urgent correction, as well as opportunities ripe for consolidating and improving the national supply system.

Movement in this latter direction in the 2000s marked a decisive shift, as investments across all aspects of resource production and transport radically boosted output. Raw coal output in Inner Mongolia, for example, rose from 72.48 million tons (Mt) in 2000 to 786.65 Mt in 2013, making the province the top producer in China – ahead even of Shanxi Province, historically the epicentre of the national coal industry. The most dramatic increases in output in Inner Mongolia were registered in Ordos Municipality, where the Heidaigou-Ha'erwusu complex is located. During the same period, output in Ordos skyrocketed from 22.9 to 576.1 Mt (peak output in 2012 was 596.7 Mt), a 25-fold increase over base year output.[3] The Heidaigou and Ha'erwusu mines during this period boosted production from 6.1 Mt to 57.6.[4]

It is interesting to note, however, that while coal output rose spectacularly in aggregate terms in Inner Mongolia, it was by no means an orderly affair that followed the prescriptions of coal industry five-year plans or the vaguer guidelines of the Open the West agenda (Woodworth, 2015). The two nationally dominant 'super-large' state-owned enterprises – Shenhua Energy Corporation, the parent company of the

Heidaigou mine operation, and China National Coal Group – received preferable loan conditions for technological upgrades, infrastructure expansion, and easy approval of production licences and finance under Beijing's regional development campaigns. But state-owned mines controlled by state entities at lower levels as well as private firms, meanwhile, continued to proliferate and prosper during the 2000s, despite episodic campaigns mounted by National Development and Reform Commission to consolidate or close them down. These latter firms, in fact, supplied the lion's share of output in the 2000s, operating in plain sight, despite flagrant regulatory violations (Figure 8.1).

The constellation of mines across Inner Mongolia, including the Heidaigou operation, thus culminated decades of intent to exploit the abundant resources in China's peripheries. In an echo of Scott's thesis of 'seeing like a state', the core principles within the relevant policy debates have been consolidation, concentration, and rationalization (1998). In practice, this has meant mass displacement to make space for mines, the introduction of enormous – and enormously destructive – new machinery and facilities, and the disembowelling of landscapes, causing destruction to pasture and fields, depleting groundwater, and decisively disrupting livelihoods in the region.[5] It is against this backdrop that protests and expressions of discontent have erupted across Inner Mongolia contesting

Figure 8.1 Machines work the coal seam in the pit of the Heidaigou coalmine, in Ordos Municipality, Inner Mongolia. Photo: Max D. Woodworth.

the expansions of mining. I will turn to these later in the chapter, but it is first necessary to survey the second landscape of the frontier, one that exhibits the afterlife of fossil-fuel production in the manic production of new city spaces.

A dispatch from the 'ghost city' square: Efforts to document the energy frontier in Inner Mongolia brought me again and again into Genghis Khan Square, a monumental civic space that fills the entire central axis of the Kangbashi New District, a new-town project initiated in 2004 by Ordos Municipality. Built during the coal boom, the new town has since become a primary exemplar of China's so-called 'ghost cities' in light of the scant residents who've moved in. But the municipal library is located on the edge of the square, and so my commute to its archives required traversing part of its enormous expanse on foot. Struck by the square's size over several years of visits, I dedicated an afternoon to walking its full length. The 'square' is, in fact, a series of squares and plazas stretching nearly two and half kilometres from north to south. Each section is designed to reflect a particular motif related to the life of the Khan or to other aspects of regional culture. However, the designs are essentially invisible from the ground level; I have seen them in satellite images and computer renderings but without any height there is no vertical perspective from which to appreciate the designs. Up close, there is little to see, other than vast marble surfaces and landscaping. Impressive, massive bronze statues and public art are scattered throughout the space. Yet on every occasion that I have traversed the square, including this instance, other visitors are nowhere to be found. The space is entirely devoid of people, save for a handful of gardeners sheltering from the intense sun amid shrubs and a lone woman tending a refreshments kiosk. On one of the squares, I come across the entrance to an underground shopping mall. Signage suggests an emporium, but the glass doors are secured with chains and a padlock, and beyond the door the marble arcade recedes into darkness. At a casual pace, pausing to document the various features of the square and to chat with the gardeners, it takes nearly two hours to cover the full length of the square. The size and emptiness of the space leaves me with an uncanny sense of strolling through a massively scaled model of a city rather than the real thing.

As studies of resource peripheries in varied settings have shown, among the economic and social reverberations of resource booms is the production of new urban spaces (Bradbury, 1979; Chapman, Plummer, and Tonts, 2015; Keough, 2015). Inner Mongolia has been no exception. Across the region, city building was pursued with great zeal amid the resource boom, and a ferocious round of urban construction thoroughly remade its urban spaces. After decades during which urban form was functionalist and compact, the spate of intensive urban development in the 2000s produced sprawling landscapes of new residential high rises, opulent office and retail complexes, massive public squares and parks, new industrial zones, and exclusive gated communities for the region's class of wealthy elites. Kangbashi New District was one such project among numerous others located across Inner Mongolia (Figure 8.2a and b).

Yet unlike historical North American resource boomtowns that expanded spatially in a frantic rhythm set by a crush of newcomers seeking their fortunes in resource extraction (Hostetter, 2011), Inner Mongolia's boomtown spaces are first and foremost city-making projects initiated by local states eager to use their control over the primary land market to emulate the speculative land development that has so dramatically remade cities across China. Low population density and vast rural spaces promised tantalizing rent gaps that became richly exploited, as new towns and development zones were opened in rapid succession in jurisdictions where mining was also expanding. Wealth generated by the boom, and a ceaseless drumbeat of local-government boosterism advanced through state-run media, bred faith in the economic potential of the region. By the mid-2000s, Ordos, for example, had rebranded itself 'China's Kuwait' and 'China's new coal capital'.

As the dispatch from Kangbashi above suggests, the region's new urban spaces present a contradictory tableau. In line with authoritarian regimes in other settings, the making of new urban spaces was an occasion to experiment with utopian forms (see Koch, 2010). Centrally planned city layouts give the new town a strict sense of geometric rationality typical among authoritarian modernist schemes, while symbolic architectural and design features saturate the spaces to produce the new city spaces as monuments to the developmentalist state. Exemplifying this tendency, Ordos' mayor expressed the notion of the Kangbashi new town as a form of material and spiritual elevation at the project's founding: 'Humanity's history is one of science overcoming ignorance and civilization replacing backwardness. The city is the hearth of civilization and its greatest achievement! We must build a big city, one that is first-rate in China, one that has connected tracks with the world, and is modern and globalized. We must use a big brush to write a big new chapter, to adopt a big vision to produce a big image'.[6] This ethos is widely shared among urban officials across Inner Mongolia (and, indeed, most of urban China) and has been translated into a proliferation of urban projects similar to Kangbashi that receive less media attention but are no less monumental, such as the Tiexi Economic Zone and Dalu New District (Ordos), Binhe New District (Baotou), Xilinhot New District (Xinlin Gol), among others.

(a)

(b)

Figure 8.2 (a) and (b) The Kangbashi New District in Ordos Municipality has developed as a site of financial speculation and grand visions of urban change on China's mining frontier. Photo: Max D. Woodworth.

On the other hand, Kangbashi and the projects just listed have fostered turbulent speculative bubbles in real estate fuelled by private investors' exuberance surrounding the resource booms. As local households and investors snatched up properties with great enthusiasm, forests of residential skyscrapers and class-A office towers rose in former desert and grassland. In Kangbashi alone, more than 300,000 residential units were completed between 2006 and 2013. Yet the new town today has an official population of less than 50,000, a figure that is almost surely inflated. Vacancy rates in some developments are around 90%, though properties are entirely sold (see Fang et al., 2015). The pursuit of the capital gains in property, legitimated by the skyrocketing home and land values through much of the 2000s, did not necessitate in-migration. New-town properties held as speculative assets sit vacant, the new towns manifesting a kind of Potemkin urbanism, where massive buildings are little more than empty shells. The eeriness of such spaces has not escaped the curious public; Kangbashi, for example, is now a popular tourist destination and has even been the setting for several professional skateboard films, as the smooth new surfaces, vast empty spaces, and azure skies are both practical and highly telegenic.

The side-by-side presence of spectacular and speculative spaces should perhaps be seen as an intrinsic feature of resource frontier boomtowns. As Ellen Hostetter (2011) observes, resource boomtowns are spaces defined by 'excess' in all domains encompassing monumental state projects, speculative ambitions, and personal consumptive behaviours in reflection of the enormous easy riches that are the perennial promise of resource bonanzas. As Inner Mongolia emerged as China's new energy frontier in the 2000s, these impulses spilled into a proliferation of eye-catching architectural icons and over-production of properties forming landscapes of supreme size and extravagance, weaving together apparently disparate and disjunctive moments into an integrated 'economy of appearances' (Tsing, 2000) premised on the frontier's radical forms of speculation and spectacle.

Discussion

While dramatically reconfiguring material environments, extraction and urbanization in Inner Mongolia have also produced a rich cultural archive that forms a complex repository of representations about the space. Kangbashi and other new-town projects across Inner Mongolia's resource frontier are now routinely represented as 'ghost towns', an appellation that has since been applied to failed or struggling megaprojects throughout China. In this sense, frontier processes have 'come home' by supplying a discursive frame that seeks to explain China's broader urbanizing and speculative excesses in the 2000s. The large and growing number of news reports, scholarly studies, photographic projects, and documentaries pertaining to the frontier's fraught urban megaprojects have made them symbols not of limitless growth and development success, as the 'China's Kuwait' motto would suggest, but rather as controversial instances of extravagance, waste, and mismanagement. Rather than regions on the rise, they are portrayed as sites of ongoing ruin, reverberating Ann Stoler's (2013) notion of 'ruination', a term that captures the continual impacts of 'imperial effects' upon places and societies caught in the zones of imperial activity, of which extraction is a signal example.

In her writings on the gigantic, Stewart observes that the figure of the giant became a frequent totem in medieval European cultures appropriated by commoners to comment critically upon structures of power (see Stewart, 1993, especially 78–86). It is interesting, in this light, that a 2015 semi-fictional documentary, which rapidly became a sensation across China through its devastating portrayals of mining, metallurgy, and city building in Inner Mongolia, was titled *Behemoth* (Ch. *Bei xi moshou*) and adapted Dante's *Divine Comedy*. The film, by the Chinese director Zhao Liang, recounts a dead man's passage through purgatory and hell, and ultimately his ascendance into paradise. The setting for this tale is the contemporary landscape of Inner Mongolia. Its protagonist, slowed by pneumoconiosis (i.e. black lung), is shown trudging through ravaged mining landscapes and cacophonous factories and smelters, finally terminating in a new city of high rises and broad boulevards curiously devoid of people. For viewers able to read the depicted street signs, the urban endpoint is unmistakable: it is Kangbashi. Notable in the film is the manner in which machines appear ludicrously big; workers, by contrast, are ant-like, scurrying about their tasks speechless from exhaustion and concentration. An invisible presence, that of a superordinate power seeming to command this hive of activity from afar, looms throughout the film and is nothing short of monstrous.

While the political economy of regional development is never named as such, *Behemoth* has been hailed as a trenchant commentary on the frontier's 'economies of violence' (Watts, 2014) and its ecological impacts. Given the popular sensation of the film and the currency of the 'ghost city' trope, such hyperbolic representations now constitute a key part of what the frontier has become as a minescape. They also suggest that more is at play in the reordering of resource frontiers than the convergence of institutional and technical conditions for extraction. The Inner Mongolian frontier coheres as a space of radical imaginings, real dislocations, and material hyper-exploitation, the scale and comprehensiveness of which compels a view of such spaces as gigantic in their material and discursive dimensions.

Notes

1 See Zeng, P. (2010). *Xibu Da Kaifa Juece Huigu [A retrospective on the policymaking for the Open the West campaign]*. Beijing: Xinhua Publishing House.
2 See Mao Zedong. 'On the Ten Major Relationships'. https://www.marxists.org/reference/archive/mao/selected-works/volume-5/mswv5_51.htm (accessed 22 April 2017).
3 Production statistics gathered through Ordos Bureau of Statistics website. See www.ordostj.gov.cn (accessed 22 April 2017).
4 See Jungar Banner Bureau of Statistics. (2012). *Zhunge'er qi guomin jingji he shehui fazhan tongji baogao (Jungar Banner Economic and Social Development Statistical Report)*. Xuejiawan, Inner Mongolia.
5 For analysis of environmental impacts of coal mining in Inner Mongolia, see Greenpeace (2012).
6 See Yun, F. (2006). Zhan zai xin qi qidian shang [Standing at a new starting point]. *Ordos Daily* (August 1).

9

Private Healthcare
in Imphal, Manipur
Liberalizing the Unruly Frontier

Duncan McDuie-Ra

Introduction

Imphal – the capital city of Manipur, a former kingdom presently located on India's far eastern frontier – has been occupied continuously since 1891 by British and Indian armed forces. Through the last seven decades Manipur has been an arena for separatist struggles by groups seeking the restoration of the formerly independent kingdom, by left-wing groups seeking independence and radical social restructuring, and by ethno-nationalist groups challenging the boundaries and legitimacy of contemporary Manipur. The latter struggles have brought members of different ethnic groups living in Manipur into conflict with one another, with India acting as provocateur and referee. Imphal sits at the geographic and political epicentre of this arena. From 1980 to 2004 the city was declared 'disturbed' by the Indian government and subject to the notorious Armed Forces Special Powers Act 1958 (AFSPA) affording legal protection to the Indian military and paramilitary. Imphal may no longer be officially disturbed, yet violence characterizes life in the city – as transacted physical violence and what Andrew Herscher and Anooradha Siddiqi refer to as 'spatial violence', the deeper and slower structural forms of violence that contour political historical categories such as 'development', 'reconstruction', 'modernity', 'peace', 'progress', and so on, which they argue is 'a constitutive dimension of architecture, urbanism, and their epistemologies' (Herscher and Siddiqi, 2014: 270–271). Military occupation, a weak civilian government, and the presence and

Frontier Assemblages: The Emergent Politics of Resource Frontiers in Asia,
First Edition. Edited by Jason Cons and Michael Eilenberg.

power of multiple armed and unarmed organizations representing different ethnic communities and territorial imaginaries (underground groups hereafter) typify what Elisabeth Dunn and Jason Cons call a 'sensitive space', where people are subject to multiple 'interwoven projects, logics, goals and anxieties of rule operating at once' (Dunn and Cons, 2014: 102).

As a frontier assemblage, the transformation in Imphal reveals both the practice of assembling, what Tania Murray Li calls 'the hard work required to draw heterogeneous elements together, forge connections between them and sustain these connections in the face of tension' (Li, 2007b: 264), and the agency of 'situated subjects who do the work of pulling together disparate elements' (Li, 2007b: 265). This particular frontier assemblage, the conjuncture of reinvention in Imphal, has varied 'situated' drivers. One the one hand, Imphal is subject to state-orchestrated plans to create new markets and better connect the frontier to the rest of India (west) and to Southeast Asia (east). Articulated in the 'Look East Policy' and *North East Vision 2020* document, Imphal is to be recalibrated as a 'gateway city' opening up the remote and recalcitrant frontier to national and global capital (McDuie-Ra, 2016; see also Choi; Woodworth, this volume). Imphal's remoteness is being transformed into a new site of productivity. As Jason Cons and Michael Eilenberg argue in the introduction to this collection, invoking Anna Tsing, the incorporation of marginal and remote areas – in both geographic and relational senses – through productive and extractive transformations, creates 'new articulations of territorial rule, accumulation, and security'. At the same time, connectivity brings Imphal and surrounding frontier space under tighter control of the state: a project implemented with limited success through decades of state violence and one now pursued through an assemblage of market expansion, state-infrastructure (including for natural resource extraction), and a ubiquitous rhetoric of 'opening up' that resonates with local and national imaginations of new India. The armed forces, both regular and paramilitary, and the extraordinary laws that protect these forces have not retreated and are a component part of the frontier's incorporation and recalibration.

On the other hand, non-state actors, in this case entrepreneurs in the health sector (and perhaps vague 'other forces' supporting them) are instigating an adjacent transformation and recalibrating the frontier in a different way. This endogenous transformation aligns with the vision of the frontier as a productive site, a zone of opportunity, manifest in the creation of a transnational health hub (see also Middleton; Günel, this volume). The hub also produces an unexpected frontier resource: trained health workers who are in demand throughout India and beyond as well as patients moving to Imphal to seek treatment from across international and internal borders. The success of the health sector reinforces separatist claims and fuels imagination about future sovereignty, perhaps the ultimate recalibration of the frontier.

There are other effects too, effects that demonstrate the ways 'rule by chance' in sensitive space is part of the assembly process beyond binaries of control and opportunity (see also Swanson, this volume). This chapter analyses these effects in Langol, a ward on the edge of the city. In Langol, land and rule are being transformed on the edge of the zone of opportunity itself. Here the two visions of the frontier – state and non-state – feed off each other and exemplify a 'transient convergence', what Jon Anderson defines as 'a temporary coming together of many "things," an ongoing assemblage of entities/processes' (Anderson, 2009: 123). The expansion of the health hub has brought areas in the peri-urban fringe under state control. Meanwhile the state connectivity project provides new opportunities for the expansion of the health sector and its sovereign claims. Both appear fragile.

With this in mind, I explore the assemblage of interests, actors, flows and ideas that are transforming a small zone within this particular frontier, Imphal's private health sector. In doing so I make a fourfold argument. First, Imphal's private health boom has produced a resource frontier that is unexpected – health infrastructure and the out-migration of health workers to other states – and assembled in a markedly different way from other cases attracting attention in the region and in this volume. Second, the private health sector in Imphal complicates the location of the state in the transformation of the frontier. In the shift from a site of quasi-imperialist control, the state is neither in deliberate retreat nor can it impose its presence in this sensitive space. Third, the growth of the private health sector generates – and also reflects – a form of sovereign power produced through medical infrastructure and articulated as self-sufficiency. It is an expression of what can be achieved independently of the Indian and Manipur state governments. Finally, the private health sector boom has produced an unexpected fillip to state control of the northwest fringe of the city. Excavation, land grabbing, and land titling have made the once anarchical zone legible, compelling past-settlers to prove their right to dwell and demonstrating the convergence of state and non-state frontier making.

This chapter has five sections. The first offers an ethnography of Langol as an entry point to Imphal and the contours of this frontier assemblage. The second provides a brief background on Imphal making a case to consider it as 'sensitive space'. The third focuses on Shija hospital, the showpiece private hospital in the city. The fourth explores the ways patient mobility demonstrates non-state control over movement through patches of territory evoking the possibilities of future sovereignty, before closing with a brief conclusion.

Walking the Frontier City

From the fourth-floor window of the Shija Private Hospital there is a fine view of the Langol-Thangmeiband road running along the foothills into the distance. In the foreground is a church built on a hillock, a cluster of houses on various flattened patches of hill, and a row of restaurants and small hotels. A sign marking the start of the 'reserved forest' created in 1933 is lost among the advertisements for *Palmei Fooding and Lodging*, *Triune Medicos*, and bill posters advertising medical tuition. A bus full of young women in bright uniforms passed on its way to the hospital's own nursing college. Down on street level, the reflective glass of the hospital's domed main building stands out against the backdrop of the surrounding foothills. Branching out from the main building are a series of single-story clinics and specialist centres, offering everything from obstetrics, to cardiology, to cosmetic surgery. Outside is a huge signboard listing the names of all the doctors who practise in these clinics. Almost all of them are from Manipur. This is significant given the brain drain out of the region usually creates the opposite scenario; professionals and administrators from India sent on hardship postings to the unruly frontier.[1]

Part of Shija's car park is given over to the hospital's own ambulance service. At the rear of the compound is a blood bank and transfusion centre funded by the Indian government's special ministry for development in the frontier. The blood bank is the only one of its kind in the region. During a 2013 visit, I toured the blood bank with a staff member from the hospital's public relations office. He mentioned that the blood supply to Shija from the public hospital was unreliable. He paused and said, 'We need a reliable supply of blood because a lot of people get shot here'. At night, while much of Imphal is in darkness, Shija glows from the lights generated by its own power supply. The site has been dubbed 'health city', a phrase that even the Chief Minister of Manipur has begun to use (Figure 9.1).

Started in 1996 in the swampy forest of Langol Lamkahi, Shija employs 600 staff and has 145 in-patient beds. It is the largest and best reputed of the 27 or so private health facilities in Imphal. Older residents remember Langol, where Shija is built, as a place of flooded terrain, trees, medicinal plants, and wild animals. It was a place rarely visited by city dwellers. Aversion to settling in Langol was not just related to unstable ground but to unstable cosmology. Langol was (and is) haunted by deviant *lai*,[2] by animal spirits, and by Imphal's deviant dead. The reserved forest was a dumping ground for casualties of the violence gripping the city from the 1980s – especially those killed by the armed forces and paramilitary in 'fake encounters'[3] (see Phanjoubam, 2005;

Figure 9.1 Advertisement for cosmetic surgery at Shija hospital featuring what appears to be a Korean woman, referencing both an aspiration for a certain kind of beauty and a country famous for cosmetic surgery. December 2012. Photo: Duncan McDuie-Ra.

Human Rights Watch, 2008; Dobhal, 2009; Hedge, 2013). Bodies that cannot be cremated in city localities are also dumped and/or burned in what is left of the forest by family members – usually in cases of suicide.

The narrative of Shija hospital's creation suggests an empty space. During time spent walking through the area on visits between 2012 and 2016 an alternative narrative of settlement emerges. Langol is where the valley meets the hills, a distinction between districts (Imphal West and Senapati) and different land use regimes. In the valley, conventional land title operates while land in the hills is under customary land tenure governed by tribal councils. The exact location of the boundary was unclear when the foothills were covered in forest and swamp. Ambiguity attracted settlers from different tribal communities. Some fled inter-ethnic conflict and operations by the security forces and underground groups. Others came to Langol for land to build a house and farm, some came to hunt, some to follow others, some to escape the authorities, some because they received a celestial message.

A short walk west from the hospital is the start of Hebron Veng Street, a dirt road stretching 200 m or so into the surrounding fields. Hebron Veng Street is named after the Hebron neighbourhood (*veng*) in Churachandpur, a district headquarters in southwest Manipur. Demands for increased autonomy for the Kuki ethnic group have shaped politics in Churachandpur in the last decade, resulting in a cycle of protests, violence, and crackdowns by the armed forces. Like many other communities in the state over the last 20 years, scores of families from Churachandpur have found themselves starting a new life for a year, two, or 20 in Imphal's peri-urban fringe (Figure 9.2).

Hebron Veng Street is lined with small houses set in neat rows made from sheets of asbestos fibre mixed with cement and woven bamboo. Most have new galvanized steel roofing brought across the border from Myanmar. A few houses have small kiosks added to the front and walls feature calendars produced by Kuki student organizations and pictures of martyrs killed by security forces. At one end is the recently completed

Figure 9.2 New settlements in Langol. The rows of buildings along the road are small shops. May 2014. Photo: Duncan McDuie-Ra.

Langol Independent Church. In parallel Spring Villa Street are the Chongtu Baptist Church, Bible Baptist Church, Kuki Baptist Church Langol, and Layman's Evangelical Fellowship.

Throughout Langol, it seems there is a church every hundred metres or so, quite a sight on the outskirts of what is believed to be – and politicized as – a city of Vaishnavite Hindus and animist Sanamahis from the dominant Meitei community and Muslim Pangals. The politics of the hill-valley divide in Manipur, namely that Meitei and Pangal communities live in the valley and Christian tribal communities in the hills (Jilangamba, 2010; Lunminthang, 2016), is challenged in areas like Langol, where hill communities have sought to establish themselves in the valley on the edge of the city. Some churches are complete and functioning. Others are half-built with a functioning room or two and a name, often of a village in the far-away hills. Wandering around one of these I met a caretaker, Peter. His church had a façade, door, and crucifix. Peter said the money for the church was coming slowly from a congregation in Chandel district. He hoped it would be finished soon and more people from the village would come and settle. It was important that this happened quickly, as the community needed to stake a claim in Langol before all the land was taken. He quickly added, 'the government will not remove a church'.

The reach of the civilian government was limited in Langol even as the area's population grew. This changed in the 2000s. After the National Games were held in Imphal in 1999[4], the houses in the village were sold off to well-connected civil servants or awarded to successful athletes. Shija acquired more land to extend its operations including housing for staff and students and an oxygen production and storage facility. The flows of people going to and from the hospital have made the routes through Langol viable for transport providers and the route now sees a stream of auto-rickshaws, converted vans, and Shija's own buses. Shops, hotels, and pharmacies followed. As vacant space in the city proper shrinks, the area has become a desired location for Imphal's professional classes, increasing the cost of land. Alongside these mansions are several drug rehabilitation homes, a yoga retreat, and a Hindu *mandir*. What was an area for those escaping trouble in the hills, unwanted attention from the authorities, or looking for better places to hunt animals is now a part of the city. What was wild and unruly has become tame and orderly.

Imphal as Sensitive Space

The total population of Imphal municipality is 414, 288 (Census of India, 2011). Imphal occupies a relatively small area, officially 34.48 km², though settlement goes well beyond municipal boundaries. Spatial control in Imphal is fragmented, a prime example of the ways non-state actors 'crosscut or superimpose themselves on the territorial jurisdiction of nation-states' (Ferguson and Gupta, 2002: 998). There is a temptation to think of the city as divided into distinct zones of control: some run by the armed forces, some by the civilian Manipur government, some by underground groups, and some where control is absent. Yet the boundaries of these zones shift constantly in response to the ways space is utilized and contested in the city. Even in sites where control appears firm, such as paramilitary barracks, bomb blasts, civic pressure to relocate, and even dissenting graffiti on the walls suggests some volatility (Figure 9.3).

Imphal was declared a 'disturbed area' in 1980 after violence plaguing the rest of the state travelled to the city. A disturbed area is any designated territory within the current (though disputed) borders of India where extraordinary laws can be enacted. Only the Ministry of Home Affairs or the Governor of the respective state can declare an area disturbed for a period of six months, though there is no limit on the renewal of disturbed status and parts of Manipur have been declared disturbed continuously since 1958. The AFSPA operates in all disturbed areas in the frontier permitting any member of the Indian military and paramilitary (armed forces hereafter) to fire 'even to the causing of death' upon

Figure 9.3   Pedestrian passing publicity for the Central Reserve Police Force posted on the side of a bunker but on the outside of a bank. December 2012. Photo: Duncan McDuie-Ra.

individuals acting in contravention of any law or order, carrying weapons (or anything capable of being used as a weapon) or assembling in a group of five or more people. Suspected persons can be detained for 24 hours, with unlimited extensions/renewals, and members of the armed forces are permitted to enter any premises without a warrant. Most significantly, the AFSPA provides legal protection (in the form of both de facto and *de jure* impunity) for members of the armed forces operating in a disturbed area (see McDuie-Ra, 2009; Mathur, 2012; Amnesty International, 2013).

Disturbed status was lifted from the Imphal valley, including Imphal city, in 2004 after mass protests following the rape and murder of a Manipuri woman, Thangjam Manorama Devi, by members of the Assam Rifles paramilitary gained national and international attention (Bora, 2010). Yet the armed forces still occupy the city, still administer various public buildings, and still have a major influence on political and social life (McDuie-Ra, 2016: 67–73). And although the Manipur government has its own police forces that are not legally bound by AFSPA, they operate within the same culture of impunity. Furthermore, between 20 and 30 underground groups are active in Manipur (IDSA, 2014). Some are armed groups fighting for secession from India, for changes to existing federal state boundaries, for territorial autonomy within Manipur, and for changes to ethnically determined affirmative action categories. Many have 'above ground' political parties, media outlets, and affiliated NGOs that engage with the government and the military on various issues – usually outside formal politics. Some are loose networks engaged in smuggling, trafficking, kidnapping, and extortion and in the murky world of Imphal's infrastructure development, contracting, racketeering, and – increasingly – social services including health. They protect as well as threaten, and for many residents this produces a mixture of fear and loyalty. This assemblage of legal pluralism and insecurity makes Imphal a frontier city from the view of the Indian state, the military, and the mainstream media, though not necessarily by residents, though the mainstream perception is changing as the frontier is recalibrated as a gateway.

Governing the development of the city is a challenge. Damage to public buildings through explosion and arson, such as the state assembly (2001, 2012), the state public library (2005), the public works department (2011), and the residence of the Chief Minister (2008), combined with the visible presence of armed forces, many under command of Delhi, depict a weak civilian government at the municipal (Imphal), district (Imphal East and Imphal West) and state (Manipur) level. Similarly, rundown and dysfunctional services escalate grievances against the civilian government among residents. Public works – such as the new state assembly building, the Bir Tikendrajit Flyover, the Manipur Film

Development Corporation complex and City Conventional Centre – intend to instil some faith in the civilian government and address, or at the very least avert attention from, urban dysfunction (Imphal Municipal Council, 2007).

Given these conditions, 'sensitive space' developed by Dunn and Cons serves as a useful way to approach Imphal. Sensitive spaces are notable 'for the multiple forms of power that abound, compete and overlap there and the forms of anxiety that they provoke for both those who are governed and those who seek to govern' (Dunn and Cons, 2014: 95). In these spaces people are 'constantly forced to transgress the bounds of projects, they erode specific sovereign projects – the techniques of sovereign power – and the claims to sovereign authority that they mark' (Dunn and Cons, 2014: 102). This constant erosion produces anxiety for actors seeking to govern sensitive space as it exposes their tenuous hold on territories they claim. This in turn leads to newer attempts to control and rule by chance – aleatory sovereignty. This typifies power in a constantly shifting landscape and helps to assemble the frontier in Imphal, especially in Langol.

As Christian Lund has argued in his work on public authority in Ghana and Niger, social life is 'constantly (re-) produced and sanctioned, not necessarily by one single body of "state," but by a variety of institutions which, in doing so, assume public authority and some character of the state' (Lund, 2006: 688). Legitimacy is a crucial part of establishing authority, and what is regarded as legitimate varies within different spatial and social contexts. Further, legitimacy shifts over time and in relation to the actors themselves and issues on which legitimate authority is sought and is thus established through conflict and negotiation (Lund, 2006: 693). Lund goes on to argue that a key component of legitimizing public authority is territorialization through 'delimitation and assertion of control' over a geographic area. This is exemplified in the 'health city' in Langol on the northwestern fringe of Imphal.

Liberalizing Sensitive Space

To grapple with the creation and longevity of Shija hospital, I met with its founder and director, Dr Singh. Dr Singh and his wife resigned from the Regional Institute of Medical Sciences (RIMS), the main public hospital in Imphal, and started a clinic at the present site in Langol. At the time, the government was clearing the forest to build the National Games Village. Dr Singh talked a lot about the early days: terrible roads, no electricity, few patients, fear of ghosts, and scorn from peers for leaving the security of the public system. Dr Singh insisted that dramatic change was necessary. It was 'enforced', he stressed. The public system was so bad that he had to start a private clinic to provide care and try to keep medical professionals in the state.

For medical professionals the private sector provides more career advancement, professional development opportunities, and access to new technology. It is seen as a meritocracy compared to the bureaucratic, conservative, and corrupt public sector. A common refrain in contemporary Imphal is that the private sector brings Manipuri doctors and other health workers, who once left the state *en masse*, 'home'. These are the most easily extracted resources from this frontier for the rest of India: health workers and medical/nursing students (McDuie-Ra, 2012: 62–63). Data from the National Rural Health Mission shows that Manipur has almost double the required number of doctors in primary health facilities, double the required number of female health workers/ assistants, as well as surplus pharmacists and technicians (National Rural Health Mission (NRHM), 2015). Building the infrastructure to keep them home or enable their return home is a crucial project, and one taken on by the private health sector in the production of its own sovereign claims. For health entrepreneurs, it is also easy to expand in the private sector, adding more staff, specialists, acquiring more equipment, and, in some cases, land. This is much slower in the public system where requests usually have to go to Delhi and then the state government.

In contrast, public hospitals face a number of challenges. RIMS was once the leading hospital in the frontier but now the facilities and technology are dated and much of the infrastructure is neglected. The site feels like a caricature of failed development: piles of rubbish, cows roaming around, broken windows here and there, peeling paint, and rooms caked with dust in winter and trampled with mud in summer. RIMS is still a teaching hospital, given the demand for medical training in Manipur, but competition from private training and from colleges outside the state have altered the profile of students studying at RIMS reproducing inequalities in opportunity and quality of training. Those who can afford to study elsewhere do.

The Jawaharlal Nehru Institute of Medical Sciences (JNIMS) at Porompat in the city's east is in slightly better shape after undergoing upgrades in 2012 and 2013. From time spent in conversation in the waiting areas and from friends in Imphal who have experience of the hospital, it appears that doctors see patients at the hospital for initial consultations and then encourage them to visit them at private facilities.

For patients, the private sector has a strong appeal as the standard of care is better, admission is faster, the equipment more advanced, the medicines (probably) not fake, and they will not be asked to make additional payments during medical procedures. Whether or not there is an enormous difference across all the forms of care is unknown, however, the *perception* is that private is far better. Some of the private hospitals offer their own reduced fees or free care for poor patients and revenue from specialist and elective treatment is used to cross-subsidize primary healthcare. Public money is used to subsidize costs for patients in the private sector, too, though not for all patients.[5]

The longer one engages in these conversations the more familiar they sound: the private sector is more efficient, more merit based, and a panacea for corruption and dysfunction. Yet the notion of the private health sector as an enforced split from the corrupt public system fails to account for the shifting and relational loci of power in sensitive space. Virtually all business in Imphal requires connections into the murky world of elites: conventional political and business elites are part of networks with figures of customary authority, high-ranking members of the military and security establishment, and members of underground groups. Rumour about silent partners affiliated with this murky world abound in almost all development projects. With such opaque structures of power, it is very difficult to prove any of these rumours and, perhaps, that is beside the point. Social life thrives on speculation about who is connected to whom, on who *really* funded what, and who protects whom. The adage that the private health sector in Imphal is clean, modern, and corruption-free when compared to the dirty, dated, and corrupt public health sector requires, at least, some scepticism.

Even distancing the sector from the state is not straightforward. There is a convergence between state and non-state, public and private, licit and illicit in the sector, a convergence particular to this frontier conjuncture. For instance, Shija hospital has received funds from the Delhi-based MDONER and NEC[6] (for the blood bank), and the Manipur government (after the hospital was established). Several state organizations also empanel it, meaning that patients employed by certain state agencies or under certain eligible schemes have the cost of their treatment subsidized. The telling inclusions in Shija's empanelled list are the governments of neighbouring federal states Mizoram and Nagaland. Eligible civil servants in these states can receive treatment at Shija, helping to build the hub. Here the state – the civilian government of Manipur and its districts and municipalities – are neither entirely absent nor able to impose a meaningful presence.

Producing Sovereign Claims

Shija – and the private health sector more broadly – generates new mobilities across borders circumventing Indian sovereign claims. Internal borders in the frontier are highly securitized and have all the trappings of international border crossings: heavy security, check posts, document checks (aimed at detecting non-Indian citizens as well as Indian citizens ineligible to settle in the border states), opposing state police forces eyeing each other, and unofficial check posts set up by underground groups to tax vehicles passing through territory they claim. At an historical conjuncture when eastward connectivity from Manipur to Southeast Asia is being celebrated as anathema to poverty and anti-India sentiments, internal borders between frontier polities remain highly sensitive and militarized, especially Manipur's shared borders with Nagaland – the site of the Mao Gate standoff in 2010 (McDuie-Ra, 2014), and with Assam – seen as the harbinger of illegal migration into Manipur (McDuie-Ra, 2016: 107–114). The emergence of the health city brings patients to Imphal from across all these borders. It is a different kind of resource frontier: a zone to visit for operations, transplants, and medical training despite the multitude of risks.

Outside each ward at Shija is a whiteboard with the name of every patient and their hometown. These include neighbourhoods in Imphal, towns elsewhere in Manipur and the frontier, and Tamu – the town on the Myanmar side of the international border. When I asked a doctor about this I was told that whenever someone has come from Myanmar, even if they come from towns and cities hundreds of kilometres within Myanmar territory, they write Tamu, as staff members in the hospital can locate Tamu and know the patient may not speak Manipuri.

Shija is pushing the Indian and Manipur governments for a visa-on-arrival system for patients from Myanmar, even if they do not have a passport. In the meantime, the hospital circumvents Indian sovereignty through its own networks. On one visit to Shija in 2013, Abocha, from the hospital's public relations, explained the process of bringing patients across the border. Agents on the Myanmar side gather the information about patients and house them in Tamu. When there are enough patients to warrant a trip, Shija sends a bus to Moreh (the Manipur side of the border) to collect the patients. Abocha dealt with my series of questions about the checkpoints on this stretch of road – staffed by police, military and paramilitary, and by underground groups, by pulling out a sheet of paper and placing it on the table in front of me. Passport photos of eight patients from Myanmar were glued on the left-hand side of the paper and their names, hometowns, and ailments were hand-written in a column to the right. Sheets of paper like this one are presented at the border, the names of patients listed, and when the patients leave back through the border these are checked off. My series of 'how' questions were met with a shrug, suggesting the system 'just worked' or the details were not for me to know. In this way, local networks facilitate alternative cross-border mobilities and the hospital produces a kind of sovereignty through control over territory and infrastructure, albeit in small patches.

There are also official cross-border medical exchanges. Doctors from Shija run health camps in Monywa, a town over 300 km inside Myanmar. Shija undertakes a 'Smile Train' mission – a travelling entourage of surgeons doing free cleft palate operations – in western Myanmar and government officials from Sagaing Region have visited Shija and discussed building a government-funded Myanmar lodge for patients. What is striking about this are the positions being performed. The doctors from Manipur, cast as a dysfunctional state and India's turbulent frontier, perform the role of technologically advanced donors extending medical care to a (perceived) disadvantaged population across the border.

Shija's slogan is 'Towards Changing the Health Landscape of Southeast Asia'. It appears on their promotional material and on billboards in Imphal and other cities in the frontier. The hospital makes no attempt to locate itself within India. It re-places itself, and to some extent Manipur, within Southeast Asia or at least at its doorstep. It does not only see that it can play a catalytic role in improving healthcare in Southeast Asia – though in practical terms this means western Myanmar. This is certainly a marketing strategy; however it is drastically different to the kinds of corporate language common in the frontier that focuses on demonstrating belonging *to* India, usually at a later phase of development suggesting goods, brands, and technology take time to arrive, all reminders of both the distance from the heartland and the place of the frontier within the nation. Shija's slogan does not bother with the nation at all.

Conclusion

What is behind the reinvention of this space of extraction and unchecked state violence (and counter-violence) as a 'zone of opportunity'? The private health sector in Imphal suggests the 'old' frontier of the colonial and postcolonial imagination, namely an unruly zone to be secured, is transforming into a new frontier for capital (see Middleton; Anderson, this volume). This signals a shift from old forms of extraction to new notions of the resource frontier; in this case the new resources are human – health workers – and infrastructure specific to healthcare. It suggests a transformation driven from below, by local entrepreneurs and professionals, rather than from above, by national and global capital or by state intervention. It suggests, in short, a kind of 'frontier liberalization' is taking place where service provision is in the hands of non-state actors tired of state dysfunction and in spite of entrenched militarization. Yet as the sector has grown there is a convergence with state visions of frontier transformation, not just the local state but also neighbouring state governments and the distant administration in Delhi. Rather than retreat, the local state appears to be latching onto the transformations taking place and finding new ways to expand. In the meantime, the over-bearing state, India, and its armed forces, persists in ordering space, mobility, and engaging in violence with impunity. Yet patients and pro-viders of care circumvent their apparent control of territory and mobility.

The assemblage of forces driving the sector is also difficult to cast. Doctors and other health professionals are pioneers of the sector's growth, yet without land, labour, and investment the sector cannot thrive. The extent of the connections between these pioneers and the murky world of elites representing various systems and networks of legitimacy and 'state-likeness' is opaque yet looms over the sector and its operations. The private health boom is also a claim by these forces for an alternative vision of sovereignty. The sector generates – and also reflects – a form of sovereign power produced through medical infra-structure and articulated as self-sufficiency. It is an expression of what can be achieved if the Indian and Manipur governments could be bypassed, if the system of neo-colonial dependency that permeates insti-tutions from Delhi to Imphal municipality could be disrupted. For a polity where the desire to reinstate sovereignty lost in 1949 has been a powerful determinant of the present conjuncture, the private health sector is a model of what can be achieved if the people of Manipur are left to their own devices. It is a symbol of both lost autonomy and future capacity for sovereign rule separate from India, albeit contained to one sector, one zone, one aspect of social and economic life.

Between these forces is the land itself and the communities who have established themselves as settlers fleeing violence, seeking opportunity, and taking advantage of ambiguity and minimal state presence. As Imphal expands towards Langol following the success of the health city and the value of the land increases, a new frontier is being made. Here settlers old and new create material artefacts of their longevity and legit-imacy, hoping to avoid eviction and expand their community before all the land is taken and the once anarchical zone on the edge of the city becomes fully legible.

Notes

1   This is an adaptation of Bengt Karlsson's (2011) phrase 'unruly hills' to refer to the hills areas of Meghalaya extended to the entire frontier region.
2   Lai are Meitei female and male deities that are sacred and numinous. They are worshipped among the Meitei community alongside Vaishnavite Hindu-ism, Christianity, and in the revived Sanamahi religion. See Parratt (1980).

3   A 'fake encounter' is a term used by human rights groups, civil society orga-
    nizations, and more recently the Indian Supreme Court to identify the tactics
    used by the armed forces to justify killing civilians in different parts of South
    Asia. An 'encounter' killing refers to killings by the armed forces in self-
    defence when they 'encounter' members of rebel groups. A 'fake encounter'
    is when the armed forces kill civilians posing no threat and then claim that
    they acted in self-defence. AFSPA aids fake encounters by preventing mem-
    bers of the armed forces from facing charges for killing civilians.

4   Hosting the 1999 National Games was a chance to demonstrate to the rest
    of India that Imphal was a modern, developed, and stable city: none of which
    were particularly evident then or now. It was a chance to be on the national
    map and for the Manipur government to perform a deeply symbolic embrace
    of the nation, one at odds with the prevailing currents of local political life.

5   Rashtriya Swasthya Bima Yojana is a national health insurance scheme
    targeting families below the poverty line. Private hospitals often receive
    patients under this scheme. See Vellakkal and Ebrahim, (2013): 24–27.

6   MDONER is the Ministry for Development of North East Region, a special-
    ist ministry unlike any other in India. NEC is the North East Council. It has
    a similar function but the Chief Ministers from each of the eight federal
    states in the region are members.

Part IV
Frontier Re(Assemblies)

Framing Essay
Framing Frontier Assemblages

Prasenjit Duara

The editors of the volume have provided us very useful descriptions of the two terms, assemblages and frontiers. Indeed, I find 'assemblage' as they define it to be one of the clearer statements in the recent literature, where the word, capacious as it is, is often used too loosely. As I understand it, an assemblage has multiple and multi-form agents engaged variously and frequently with different goals, but in the shadow of a determinate project – in this case, the expansion of the territorial state into the frontier for control and accumulation. This description works well to integrate the chapters in this section. More significantly, it reveals processes that converged across different frontier regions of Asia in over a century of colonial and postcolonial rule.

In their threefold framing of the term, 'frontier' is *imagined* as necessary for resource extraction and capitalist (as well as socialist) expansion; as zones of lawlessness and heightened security; and as wild and 'empty' spaces for exploitation and experimentation. In addition, I believe the following chapters might be grasped by considering the historical transfigurations of these three features in the modern frontier zones of East, South and Southeast Asia in the mode of convergent comparison. Convergent comparisons refer to the ways in which circulatory global forces are institutionalized in different societies. Historical forces have circulated – transmuting as they circulate – since the rise of empires in Eurasia; but the pace of circulation has accelerated since the mid-nineteenth century. What I call the *zone of convergence* is the impact of

Frontier Assemblages: The Emergent Politics of Resource Frontiers in Asia,
First Edition. Edited by Jason Cons and Michael Eilenberg.

circulatory forces that demand a response; the various sub-national and national responses, in turn, form the *basis of convergent comparison.*

Much of this vast region experienced the transformation from colonial or semi-colonial rule to independence and postcolonial nation-states during the middle of the twentieth century. In each of these societies, frontier zones in pre-colonial times were essentially physically impenetrable and/or difficult to govern. As James Scott citing Braudel has said, 'civilizations can't climb hills' (Scott, 2009: 20) (or we might add, penetrate malarial jungles). These zones were, of course, frontiers only in relation to imperial centrality, for instance Qing or Mogul, or kingly rule. They began at the edges of imperial administrative systems; they were barbaric spaces that had to be kept at bay by informal mechanisms or as buffer zones. To be sure, these frontiers were organized according to their own principles of governance and hierarchy that were relatively illegible to the older as well as modern states.

As is well-known, the modern colonial state in partnership with capitalist enterprise were able to penetrate these spaces principally through advances in military and infrastructural technologies in the assemblage. Nonetheless, they too, like their predecessors, recognized the largely ungovernable dimensions of these zones and evolved modes of colonial governance that were often quite different from colonial governmentality in the more settled areas. These modes involved the setting up of extractive industries whether in monocultural plantations, animal products or minerals while devolving autonomous governance mechanisms in these areas. They were also often segregated and cordoned off from settled populations. Cultural hegemony of the ruling power was exercised less through formal institutions than informal agents such as missionaries and cultural brokers. In a highly simplified way, we may consider these elements to represent the colonial assemblage in the frontier. This colonial assemblage would be transfigured but still come to haunt the postcolonial nation state in Asia.

The two most powerful empires by the early twentieth century in this area were the British and the Japanese, while the French and the Dutch represented more limited zones of control. Strikingly, despite the different histories of colonizer and frontier in each place, there are comparable periods and types of changes that correspond to the circulatory imaginaries and imperatives of the modern state, both colonial and postcolonial. Comparing British imperial policies of isolation and segregation of tribal populations in northeast India from the second half of the nineteenth century with the later Japanese imperial practices of segregation in Taiwan and Manchuria in the early twentieth century we see how colonial territorial projects developed into comparable assemblages shaped by common state imperatives as much as by direct and indirect influences particularly upon the Japanese who were careful students of European empire-making.

In northeast India which the British broadly called Assam, they initially clashed frequently with tribal populations without much military success in the effort to impose colonial rule over these lands. While the early purpose of the conquest was to gain access to Yunnan, Tibet and Chinese tea and markets, the northeast itself was developed into a major tea producing plantation zone. Subsequently, oil was discovered in the region and developed into a major industry. By the 1870s, the British decided to isolate the tribal regions through a policy called the Inner Line which prohibited other subjects from living or moving into these 'frontier tracts'. This policy continued to evolve militarily, politically, and ideologically, engaging local chiefs, middlemen, missionaries, ethnographers among others and eventually fused into a global discourse of 'indigeneity' as a major factor in frontier assemblage.

Japanese imperial expansion began at 'home' with the frontier territory of Hokkaido (and later, Okinawa) whose surviving native inhabitants, the Ainu, remained second-class citizens until well into the twentieth century. But it was colonial expansion initially into Taiwan where similar colonial imperatives of territorial control of resources – camphor in this case – from the highlands inhabited by the native people led to major military clashes with them and their unsuccessful 'pacification'. From the early twentieth century, the policy of 'aborigine administration' shifted from naked military suppression to devolution and co-optation and a permit-system was required for outsiders to enter indigenous territories. Meanwhile, camphor plantations continued to expand in the region by physically cordoning off the land from indigenous usage. At the same time, the indigenous population was ethnically and culturally marked out from the settled Chinese population and romanticized among the population in Japan as the 'authentic indigene'.

Similar policies were undertaken in one of the last East Asian 'frontiers' of Manchuria where the Japanese military installed the puppet regime of Manchukuo (1932–1945). By the 1930s, the Japanese colonial assemblage in the frontier had already developed a mature theory of indigeneity through which the relatively minute populations of Tungus tribes, including Manchus and some Mongol groups, came to be either territorially and/or ethnically segregated from the vast majority of the Chinese population who had been settling in the region mostly since the nineteenth century. Manchuria was envisioned classically as a largely empty frontier space (despite the settled population) and a laboratory for experiments of modernity that were considered to have been impossible in Japan.

By this time, the Japanese discourse of indigeneity became braided with the theory of nationality and indeed the Japanese (as also Chinese) words for nationality and ethnography are etymologically identical, *minzoku* and *minzokugaku*, respectively, and translatable as ethnos and ethnography. By building the narrative of Tungusic identity – of which they claimed to be the most advanced representatives – the Japanese staked their claim to protect these lands even while stripping it of its forests and minerals. As in British India and elsewhere, anthropologists who frequently harboured intentions that were more sympathetic and at an angle from colonial rulers, contributed to the colonial assemblage of indigeneity and distinctness of the population thereby contributing to what later nationalists among the settled populations would call 'the policy of divide and rule'.

These comments are intended to serve as the historical background for the following chapters which deal with the ways in which colonial assemblages are transformed in postcolonial states. The concept of assemblage allows us to see how the layering and imbrication of agents, intentions, and unexpected consequences come apart and recombine with new factors and forces in the contemporary period. Although natural resource extraction remained the goal and legacy of the frontier assemblage from colonial to postcolonial times, its most important transfiguration was the rise of ethno-national consciousness from the colonial-period discourse of indigeneity and its fusion with the identity politics of sovereign claims to land and purity of culture.

In the Darjeeling region's tea and cinchona plantation economies, classic products of the resource frontier assemblage discussed by Middleton, have stagnated as they have become riddled with labour strikes, sabotage and foot-dragging. While it is not the only reason, the sub-national agitations by the Gorkha people in this part of north Bengal have played an important part in this decline. Although Middleton is optimistic about the revival of this frontier from ruination to new life, we should also consider how, between the colonial frontier and the national frontier, identity politics have appeared as a new complicating factor. Without being necessarily pessimistic, let us remember that the entire northeast region of India is handicapped by virtue of its historical condition as a resource frontier. The new politics of rights and claims on resources in this mosaic is involuting rather than homogenizing swathes of a population as happened in more classic forms of nationalism. This has escalated violence both with the Indian state and with each other. Identity politics in the zomia terrain has transformed the landscape into something like a shatter zone.

In Christian Lentz's study of the dissimilar political conditions of postcolonial Vietnam controlled by a mobilizational and authoritarian party, the northwest highland peoples appear to be differently handicapped, but it is similarly conditioned by the frontier assemblages and their imperatives. Here, deliberate campaigns to erase highland memories and replace them by the lowland Kinh memorialization of revolutionary heroism are not the only force to marginalize the native Hmong, Khmu, and Dao swidden cultivators. The cascade of dams culminating in the mammoth Son La Dam completed in 2012, the acme of the revolutionary assemblage, has displaced tens of thousands of people and exacerbated problems of poverty and ethnicity. From innumerable such instances, it is imaginable that the latest round of neoliberal frontier assemblage is capable of not just erasing memories or containing populations but vaporizing landscapes and peoples.

10

Frontier 2.0

The Recursive Lives and Death of Cinchona in Darjeeling

Townsend Middleton

Introduction

Near Darjeeling's legendary tea plantations grows a crop with a different imperial history: *cinchona*, the miraculous 'fever tree' that produces quinine, the principal medicine for malaria throughout the British colonial period. Cinchona's arrival in India from South America in the nineteenth century thrust the remote Darjeeling Hills to the centre of a broader geography of colonial medicine and power. Yet this frontier's place in world history was always a tenuous one. The market for Indian quinine crashed in the twentieth century with the advent of synthetic anti-malarials and development of other, cheaper cinchona frontiers elsewhere, leaving Darjeeling's cinchona plantations – and the roughly 50,000 people that live on them – in the lurch. Darjeeling's cinchona plantations still exist, albeit in a dilapidated and increasingly controversial state. Cinchona's future is uncertain, as the government, ethnic groups, labour unions, and private investors wrestle for control of these resource-rich lands. Per one imaginary, the plantations are to be diversified into profitable industries. Per another, they are to become ethnically autonomous territory. Per another, these 'waste lands' represent an untapped horizon for eco-tourism. Amid these politics of repurposing, the remains of one frontier have become the grounds for a second frontier – Frontier 2.0.

This chapter chronicles cinchona's history in Darjeeling, tracing the makings, unmakings, and possible remakings of this medical frontier.

Frontier Assemblages: The Emergent Politics of Resource Frontiers in Asia,
First Edition. Edited by Jason Cons and Michael Eilenberg.

Cinchona's lives, 'death', and imminent rebirth challenge the framings of frontiers as one-and-done formations – born of boom, exhausted of life, and left for dead (see also Swanson; Lentz, this volume). Challenging this standard narrative, I instead want to theorize frontiers as *recursive assemblages*, prone to periodic (often highly contingent) cycles of birth, demise, and reformation. *Recursive assemblages* invite a rethinking of frontiers' ontology and temporality. India's cinchona project came to – and disappeared from – the stage of world history through a complex array of technological, material, human, and nonhuman forces. Botany, chemistry, land, labour, logistics, humanitarianism, and imperialism were all integral to its formation (see also Zee, this volume). Yet at no point was this a stable assemblage, but rather one where the heterogeneous elements articulated with one another in deeply contingent ways. As an assemblage, these delicate articulations would prove the cinchona frontier's greatest source of vitality and vulnerability.

Temporally, what interests me is the recursivity through which the remains of a dying frontier may be reassembled anew (see also Rubinov, this volume). Channelling Foucault, Ann Stoler has described recursive histories as those 'that *fold back on themselves*, and in that refolding, reveal new surfaces and new plains' (Stoler, 2016: 26). Her discussion of *imperial debris* subsequently explores the toxic constraints that imperial histories bequeath to the present. Like Stoler, my sketch of cinchona examines how the remnants of a seemingly bygone imperialism condition contemporary life. But moving beyond a morally unambiguous paradigm of ruination, I am equally interested in the life and futures that lurk among these remains. As an assemblage, the cinchona frontier of the Darjeeling Hills[1] has not so much died as fallen apart, leaving scattered remains. For the parties looking to repurpose these remains, the detritus of this erstwhile medical frontier represents the grounds from which any twenty-first-century alternative must be forged. Recursivity has thus emerged as a central axiom of these politics of repurposing – and a vital analytic for tracking the uncanny lives and ostensible 'deaths' of cinchona and other frontiers.

Thinking with *recursive assemblages* has particular stakes at Asia's margins, spaces that have periodically slipped into and out of focus as frontier zones. It trains attention to the global conjunctures and capitalist projects through which 'peripheral' areas of Asia have been – and remain – subjected to frontierization (cf. Tsing, 2003; Ardener, 2012). Conversely, it likewise sheds light on the dynamics of exhaustion and abandonment through which frontiers fade from view. The story of these remote areas, however, is seldom finished. Cautioning against narratives of closure, *recursive assemblages* calls for continual interrogation of the processes through which frontiers come into and out of existence. To that end, let me turn to the interplay of forces – imperial, botanical, long-distance, and otherwise – that first made this remote corner of India a centrepiece of a colonial qua medical world order.

The Makings of a Medicinal Frontier

Like many boom crops, cinchona originated a notable distance from where it was put into industrial cultivation. Cinchona hails from the Andean highlands of South America. Spanish Jesuits first 'discovered' the fever-reducing properties of its bark from indigenous Peruvians in the seventeenth century.[2] The Jesuits brought the bark to Europe, where its powder traded under names like 'Peruvian Bark' and 'Jesuit's Powder' for two centuries. The European medical establishment was initially wary of cinchona. The bark cured certain fevers, but not others. At this point, the science of cinchona and malaria were not yet known. It was not until cinchona's key alkaloids were isolated in 1820 and the malarial parasite was discovered in 1880 that cinchona's science came into view. Until that conjunction of botany, chemistry, and epidemiology, cinchona remained a medical mystery. Nevertheless, the bark's medicinal efficacy was enough to fuel a global cinchona trade, with South America supplying European markets with the life-saving febrifuge (Brockway, 1979/2002: 124–127).

With cinchona and malaria's aetiology coming into focus, the British and Dutch sought to put it to colonial use. By the 1850s, experiments were underway: for the Dutch at the Buitenzorg Gardens in Java; for the British, at the Royal Botanic Gardens at Kew (England) and its Indian proxies, the Botanic Gardens of Calcutta and Ootacamund (Madras) respectively. Things did not go smoothly. Plants and seeds from South America transshipped through Kew suffered terribly amidst the journey and heat of India. The initial experiments met with considerable failure, including a botched attempt to grow cinchona in Darjeeling in 1854–1855.[3]

Following the Rebellion of 1857, the British redoubled their efforts to cultivate cinchona. In 1859, Kew began deploying seed collectors to South America to appropriate the best strains of the fabled 'fever tree'. And appropriate they did, relying on subterfuge to smuggle seeds and live plants past the Peruvian, Bolivian, and Ecuadorian authorities.[4] Shipped through Kew to Calcutta and Madras, the seeds eventually took root in the latter, making the Nilgiri Hills the first site of successful cinchona cultivation in India (Veale, 2010).

The imperial designs of colonial botany are undeniable. Lucile Brockway has persuasively argued botany – and cinchona specifically – to be an arm of imperial domination (1979/2002). As a form of power in its own right, the ability to fight malaria was vital to British military and commercial interests. However, the British also framed the cinchona project as a humanitarian endeavour (cf. Veale, 2010: 140). The Secretary of State to India explained in 1875: 'The object has been emphatically stated by my predecessors not to be a commercial object, but one having reference solely to the supply of a cheap febrifuge to the people of India'.[5] Throughout the history of the project, this humanitarian impulse would remain in tension with its unstated imperial designs, influencing everything from the species of cinchona planted and forms of medicine made, to the distribution and marketing of quinine throughout India and the colonies.

The Darjeeling Hills were not originally high on the list of potential sites for cinchona cultivation. But Dr Thomas Anderson, Superintendent of the Royal Botanic Gardens of Calcutta, was convinced that cinchona could thrive in montane Darjeeling. In 1861, Anderson travelled to Ootacamund and Dutch Java to gather knowledge and seeds to start an experimental nursery. The government granted Anderson a modest budget and 5,000 acres of land, which he used to establish the Government Cinchona Plantations 12 miles outside of Darjeeling Town in 1862, at a site known as Rungbee (later called Mungpoo).[6] Initially, Anderson struggled with cinchona's botany. Each cinchona species required particular combinations of soil composition, moisture, sun, and elevation. Anderson shot high in some places, low in others, losing 'plants by the thousands'.[7] Gradually though, he and his staff discovered the conjuncture of botany, geology, meteorology, and labour to bring the cinchona frontier into formation – here depicted in a hand-drawn map from 1870 (Figure 10.1).

Elevation was crucial. Because each varietal preferred a specific elevational band – a range made narrower by its preferences for soil, sun, etc. – cinchona could never achieve the crop coverage of, say, tea. Certainly not in the precipitous topography of sub-Himalayan Darjeeling, where elevations quickly range from 400 to 7,000 ft above sea level. Cinchona was thus a frontier in 3D – a vertical frontier – comprised of patchy cultivations etched into steep hillsides, where ponies and porters were the only means of transport.

Developing alongside Darjeeling's tea industry, cinchona emerged as a frontier within a frontier. Like other secondary and tertiary frontiers scattered throughout Asia, it necessarily competed for space and resources with an already-established boom commodity (tea). Thus, as cinchona took root, claiming land and labour, so began a competitive and contrastive relationship – the battles of which continue to this day.

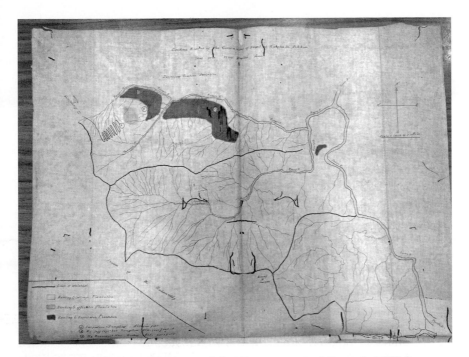

Figure 10.1   Cinchona Reserve of the Government of Bengal, 1870.[8]
Photo: Townsend Middleton.

Cinchona's botanic materiality profoundly influenced its human dimensions. Because the plant could only cover a small fraction of the land allotted to it, this meant there was land to spare within the cinchona plantations. The government could therefore provide housing and ample land to labourers, often doling out acres at a time. The promise of a comfortable peasant existence to supplement wage labour on the plantations was a powerful draw – and a stark contrast to neighbouring tea estates, where land for labourers was scarce (Besky, 2013). Like Darjeeling's tea industry, cinchona drew its labourers from the surrounding hills of British India, Nepal, Bhutan, and Sikkim. Many came straight from the private tea estates, where benefits paled in comparison to those of the Government Cinchona Plantations.

Having answered the questions of botany, geology, and labour, Anderson was on his way to proving cinchona's viability in Darjeeling. What he needed now was land – and lots of it. Darjeeling's best lands had already fallen into the hands of private tea planters. But two factors shifted the equation of primitive accumulation in Anderson's favour. First, in 1864, Anderson became Bengal's first Conservator of Forests (in addition to his duties as Superintendent of Calcutta's Royal Botanic

Gardens and In-Charge of Cinchona Cultivation in Bengal). Under his watch, the original 5,000 acres set out for cinchona mysteriously jumped to 37,000 – a transfer for which this bureaucrat-of-many-hats left no paper trail.[9] Second, at the conclusion of the British-Bhutan War in 1865, British India annexed a vast swath of territory east of the Teesta River, in what is now the Kalimpong District. With Anderson keeping watch, much of this land was declared a Forest Reserve.

The cinchona frontier here followed a familiar script. By designating this tract a Forest Reserve, the colonial state stamped its sovereignty onto this 'wild' country, conveniently eliding the native cultivators (*ryots*) who inhabited it. The area came into view as a relatively unpopulated frontier – a 'forest', as it were, ready for capitalist transformation under the eye of the British Raj. The designation was strategic in multiple ways. Most immediately, it prevented the unbridled expansion of the tea industry into the Kalimpong Hills, thereby ensuring the cinchona frontier ample room to grow when the time came. In the meantime, this newly acquired territory could produce another vital commodity for the empire: timber.[10]

The Forest Reserve designation proved a masterstroke in the politics of remoteness and that attended the making of the cinchona frontier. First, note how the value of these remote, ostensibly 'uninhabited' lands laid not so much in what they were, but rather what they could be. Figured at the edge of colonial space and time (Tsing, 2003: 5100), these remote stretches came into view as the future solution to the empire's escalating medical needs. 'Wild', 'pristine', and perfect for cinchona, these were spaces of possibility, awaiting colonial transformation. Second, note how the establishment of the Forest Reserve effectively selected – and facilitated – one form of primitive accumulation (governmental) over another (private). Declaring this land the domain of the colonial state (itself bearing a suite of imperial, humanitarian, and public interests), the government held the private capital of tea at bay. In one provident stroke, the British Raj thus created the quite-literal grounds of an emergent biopower – here condensed in the alkaloid rich, life-saving bark of cinchona. Such were the makings of the cinchona frontier – a frontier within a frontier.

The cinchona project hinged on an efficient progression from cultivation to medical manufacture. Enter chemistry. Across Europe, the extraction of cinchona's alkaloids – quinine, cinchonidine, quinidine and cinchonine – was an evolving science. The remote outpost of Mungpoo soon emerged as a key site in this nascent pharmaceutical network. Where European chemists focused on the extraction of pure quinine from cinchona's bark, the early experiments at Mungpoo focused on producing a cruder, cheaper medicine called 'cinchona febrifuge' – a powder wherein all four alkaloids were present.[11] Government Quinologist C.H. Wood combined sophisticated chemical reactions with a ramshackle apparatus of old beer-barrels and bamboo flow channels to devise this cheap medicine. Wood's process clicked with the cinchona varietals growing in Darjeeling and the project's humanitarian aims. The plantations were at this time dominated by the Red Bark species *Cinchona succirubra*. As compared to the Yellow Bark varietals preferred by quinine makers in Europe, Red Bark was low in quinine but high in other febrifuge alkaloids. Red Bark moreover thrived in Darjeeling and lent itself to simple manufacture. With the botany, geology, and chemistry clicking, the British had found a plant and a process capable of delivering cheap febrifuge to the masses.

The Government of India's first quinine factory opened in Mungpoo in 1875, producing affordable cinchona febrifuge. The pharmaceutical emphasis, however, soon shifted to the purer and more expensive quinine sulfate. This pushed the life-saving drug beyond the buying capacity of much of India and prompted a sea change on the plantations from Red to Yellow bark varietals.

The British did not necessarily forsake their humanitarian commitments, however. In the 1890s, the government launched a massive distribution scheme known as the 'pice-packet' system. Quinine made in Mungpoo would be shipped to Kolkata, then packaged by prisoners at the Alipur Jail into single-dose packets, which were then distributed to post-offices and dispensaries across Bengal, and sold at the nominal price of 1 pice.[12] The pice-packet system quickly spread from Bengal throughout India. The escalating demands put considerable strain on the factories and frontiers in the Darjeeling and Nilgiri Hills – both of which, despite their remoteness, were now central to a new geography of colonial biopower. The global and national demands for quinine radically reworked the lives and lands of these remote frontier zones. But just as these long-distance forces had amplified local effects, developments on the frontier – be they technological advancements in the lab, crop failures in the field, or labour problems in the factories – also assumed the potential to profoundly impact the national and global quinine market. Such were the dialectics of remoteness that shaped the cinchona frontier's formation (cf. Tsing, 1993; Ardener, 2012; Harms and Hussain, 2014). That life hung in the balance did much to shape these remote areas – not to mention the forms of rule, anxiety, and care that guided their transformation into a medical frontier (see also McDuie-Ra, this volume).

Heeding cinchona's humanitarian designs, the pice-packet system introduced a raft of challenges for a medical state attempting to defeat the most prolific of enemies: the Anopheles mosquito and its malarial

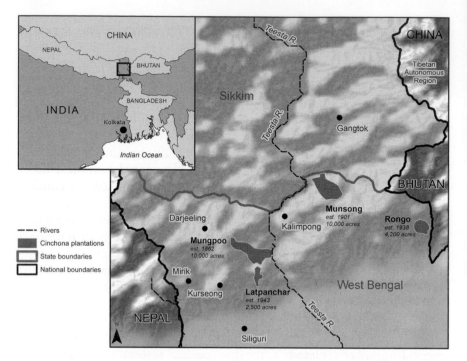

Figure 10.2   The extension of the Cinchona Frontier. *Source*: © Townsend Middleton. Map prepared by Philip McDaniel (UNC Libraries).

Plasmodium parasite. To meet the soaring demands for pice-packets, Bengal's cinchona frontier soon crossed the Teesta River and climbed into the Kalimpong Hills with the establishment of the Munsong Plantation in 1901. Still, there was a sense that it would never be enough (Figure 10.2).

Tenuous Articulations

As a strictly governmental venture, the Bengal frontier occupied a precarious niche. Private planters and cartels dominated the global cinchona market: particularly, the Dutch who produced 80% of the world's cinchona from their plantations in Java (Goss, 2014; van der Hoogte and Pieters, 2015). Ceylon planters tried cinchona in the 1870s when their coffee suffered a blight. But Ceylon's cinchona boom went bust, when the glut of bark drove prices down, pushing planter capital into another boom crop: tea.[13] Even cinchona production in India's Nilgiri Hills was a mix of government and private plantations, leaving the latter exposed to foreign market fluctuations and eventually spelling the private side of the industry's decline (Veale, 2010: 182). But Bengal was different. Precisely as a government industry not geared toward profit, the cinchona project was meant to shield the British empire from the vicissitudes of the global market. The goal was to expand the frontier to the point where India could produce enough quinine to meet its own immense demands – and perhaps too, that of the empire writ large. This would ensure a steady supply of life-saving quinine no matter what the circumstance of the global market. The aim, in short, was medicinal resource autonomy.

Confirming the British empire's vulnerability, in 1913, the Dutch government, planters, and manufacturers formed the Kina Cartel, leveraging their 80% market share to control the global quinine trade.[14] Unless and until India could produce enough quinine to meet the empire's demand, the British would have to rely on the Dutch Kina Cartel to make up the difference. The mounting imperative of resource autonomy translated directly onto the steep slopes of the Bengal frontier, where 'extension' became the order of the day. As jungles were cleared and cinchona cultivation was extended to the most remote reaches of the Darjeeling and Kalimpong Hills, the frontier came into new articulation with the global quinine market.

World War I (WWI) underscored the British empire's medical vulnerability. With supply lines from Europe disrupted, India saw its quinine reserves rapidly depleted. The government turned to Java for an emergency purchase of 1,000,000 lbs of bark, which the Kina Cartel sold at an exorbitant price. The dependency caused anxiety for British colonial administrators, who begrudgingly went through with the purchase, conceding that quinine was 'more important than rupees'.[15] WWI showed the goal of medicinal resource autonomy to be a distant possibility. British administrators estimated that in 1917 the cinchona project was meeting only one third of India's demands, let alone 'the requirements of the rest of the Empire and the allies'. Forecasting a threefold increase in demand over the coming decade, they further estimated that achieving resource autonomy would require the number of acres under cultivation to jump from the current 4,000 acres to over 50,000 acres – a massive increase.[16]

But the cinchona frontier did not lend itself to sudden expansion. Suitable land was scarce. Saplings, moreover, took eight years before yielding ample quinine for manufacture. Materially, the frontier could therefore only be but so responsive. Emerging from WWI, the government was keen to find new frontier lands. Burma emerged as first on the list. After repeated expeditions and failed experiments in the Burmese hinterlands, the British eventually established provisional plantations in the Tavoy district of southeastern Burma.[17] But unlike in Darjeeling, where the adjacent tea industry afforded labour, logistics, and infrastructure that also supported cinchona, in Burma labour was scarce, logistics were daunting, and the weather was exceedingly harsh. The Burma frontier was in these ways *too* remote. By the late 1930s, its plantations had returned to jungle.

In Bengal, the relationships between cinchona cultivation, production, and the quinine market required constant management. The government placed strict import and export controls in an effort to maintain the project's unique niche. There was constant tinkering with the cultivation and manufacturing methods to increase extraction rates. The chemistry was improving, but the Bengal frontier itself was tiring, as repeated plantings exhausted the soil of nutrients. The diminishing plant energy forced more and more acreage into cultivation – a signature paradox of frontier decline. Heeding the biomedical needs of the empire, the government expanded the Bengal frontier, opening new plantations at Rongo in 1938 and Latpanchar in 1943.

For a frontier of precarious assemblage, World War II (WWII) was a tipping point. With Dutch markets cut off by the Germans, Allied forces placed heavy demands on India's quinine stores.[18] Seeing its civilian supplies plummeting, India's states scrambled for quinine to stock their hospitals and dispensaries. The price of Javanese quinine meanwhile sky-rocketed, putting cheap government quinine in high demand. To combat hoarding and profiteering, the colonial state seized the quinine market, implementing strict distribution, pricing, and rationing schemes.[19] As the government prioritized military over civilian demands, the cinchona project's imperial underpinnings were laid bare. Humanitarianism had its limits.

Amid these wartime reckonings, a new factor entered the assemblage in the 1940s: synthetic anti-malarials. The appearance of synthetic drugs like Chloroquine and Atabrine (Mepacrine) immediately altered the biopolitical calculus. British and Indian health officials quickly recognized the long-term impacts of these new drugs. The prognosis directly impacted the outlook for the cinchona project. Suffice it to say: synthetics entered the cinchona assemblage as a necessarily destabilizing force, at once upsetting its precarious balances and changing how its elements articulated with one another. And so began the unravelling of a once-vital medical frontier.

Cinchona's Unmakings

Metaphors of 'boom and bust' are too temporally neat to capture cinchona's unmakings over the second half of the twentieth century. Better 'a death by a thousand cuts'. Tracking this slow demise requires setting the analytic of assemblage into motion. Cinchona's troubled history over this period shows that assemblages do not disarticulate or disappear all at once. Their elements remain – often as remains – working to various, if diminished, ends.

After India's independence in 1947, the cinchona project lacked the urgency of the colonial period. The war against malaria was gaining ground through insecticides like DDT and other mosquito-eradication technologies. Synthetic drugs meanwhile were cutting into the demand for quinine. By the mid-1950s, India's cinchona industry had entered a pronounced slump. Labour curtailment programs on the plantations began soon after, with the government offering severance packages to decrease its expenditures. Those who took the deal necessarily left the plantations, never to return. But most stayed, preferring to hold on to the land and livelihoods that went along with their government employment.

Where extending the frontier had been the colonial order of the day, no new acreage was put into cultivation between 1955 and 1965. With the cinchona market looking uncertain, the government began hedging its bets by diversifying into crops like tung and ipecac – and later, coffee and rubber. After the stasis and shrinkage of the 1950s and 1960s, an uptick in global demand prompted modest expansion of the plantations in the 1970s and early 1980s. This was a period of questionable optimism, however. Former managers explained this as a time of disconnect between the cultivation, production, and marketing sectors of the industry. Orders were coming down from the top to plant, but while bark was being produced, manufacturing and sales were not evolving. So 'we were just dumping the bark', managers explained. The daily work did, however, provide continuity to life on the plantations, particularly for the labourers. Not coincidentally, this was when trade unions arose to protect labour from the precarities of a dying industry.

Suffering the turbulence of the global market, the cinchona assemblage also crumbled internally – most notably during a violent subnationalist agitation in the 1980s. From 1986 to 1988, the region's Nepali-speaking (Gorkha) peoples fought to separate from West Bengal in order to establish their own state of Gorkhaland in India. The cinchona plantations – themselves a Government of West Bengal enterprise – became a frontline of the struggle. Strikes, violence, sabotage, and foot-dragging undermined the hierarchies and discipline (*anusashan*) that had been the plantations' mainstay since the colonial era. This led to an erosion of work-culture and productivity from which the plantations, by all counts, have never recovered (Figure 10.3).

Figure 10.3   Aging cinchona bark in Mungpoo, 2017. Photo: Townsend Middleton.

Signalling the frontier's demise, the Mungpoo factory officially closed in 2000. Huge stands of cinchona were subsequently left to the wild – a state in which their bark naturally loses potency. With neither much market for raw bark nor marketing schemes to move it, the little bark that was produced was left to rot in dilapidated go-downs.

Anticipating only further ruination, West Bengal has since allowed the plantations' infrastructure to slide into disrepair (see also Paprocki, this volume). Cinchona's labourers, meanwhile, continue to drag themselves to work to ensure their right to government wages and land. But with little work to do once the work-bell sounds and little buy-in to the discipline that once made the industry hum, labour has become an art of doing the bare minimum. Such are the habitations of a bygone industry.

Like its makings, cinchona's unmakings have unfolded on multiple registers: botanical, chemical, infrastructural, and socio-political. As an assemblage, cinchona's descent into ruination was not quick, but rather a slow demise through disarticulation – a falling apart, a leaving of remains. Insofar as the 'death' of a frontier figures amid these imperial ruins (Stoler, 2016: 336–378), it does so not as an instant loss of life, but a gradual leaching of vitality on myriad fronts.

A Frontier's Remaking?

Crucially, the cinchona frontier is not yet dead. Until and unless its death certificate is signed, the question remains: How might the lands and lives of cinchona be repurposed? How might the remains of one frontier become the makings of another? In these prospects of a second frontier – Frontier 2.0 – moves a recursivity that warrants some open-ended attention.

India's economic liberalization in the 1990s opened many government industries to privatization. It did not take long before suitors came calling on cinchona. The most notable – and remembered – of these was Hindustan Lever, a subsidiary of the transnational Unilever corporation. With consultation from PriceWaterhouseCoopers, Hindustan Lever began a multi-year courtship of the cinchonas throughout the early and mid-1990s. The corporation proposed a 51:49 public-private partnership (PPP) that would see the plantations changed over from cinchona to tea.[20] Trumpeting familiar PPP logics, company representatives promised development, employment, schools, hospitals, and all the rest. But the trade unions were having none of it. When cinchona workers looked across to Darjeeling's private tea estates, they saw unabashed exploitation. Land, housing, salary, and other benefits provided by the private tea industry paled in comparison to that provided by the Government Cinchona Plantations. Making their resistance known, the trade unions staged a dramatic walk-out at a public meeting of stake-holders in 1995. The situation escalated, eventually coming to a head in Mungpoo in 1996 when a 100-strong mob ambushed a meeting between PriceWaterhouseCoopers representatives and the Cinchona Directorate, throwing furniture from the offices and seizing the file containing the privatization proposal. The deal collapsed soon thereafter – marking an early victory in the fight against private capital.

The spectre of privatization re-appeared several years later when West Bengal hired the foreign consultancy McKinsey & Company to investigate ways to make the plantations a profitable public-private venture. When surveyors appeared on the plantations in the early 2000s, the trade unions mobilized, forcing the government to shelve the proposal. Another important victory.

The West Bengal government consequently finds itself in a bind. The plantations are haemorrhaging state funds, yet they also host a significant voter block. The industry remains on government-subsidized life-support, and there is a disconcerting sense that the government could pull the plug at any time. To date though, politicians have been loath to propose radical transformations, especially in light of the recent political volatility and resistance to privatization. The resources of this erstwhile frontier – with thousands of acres of prime land, abundant labour, and (dilapidated) infrastructure already in place – nevertheless remain an enticing prize for politicians and developers alike. Unlike those of the past, *this* politics of remoteness, however, must recognize and respect this frontier's accumulated histories – embodied, as they are, in the material remains and actual communities of the cinchona plantations.

With these communities having made their resistance to wholesale change known, recent proposals have been more piecemeal, with private developers eyeing eco-tourism projects, hoteliers pitching mixes of horticulture and hospitality, and, most colourfully, the corporate-Hindu guru, Ramdev, angling to build a 50-acre ashram in one of the plantation's lush valleys. Led by the trade unions, the communities of the cinchonas have stopped all of these proposals. On other fronts though, they have lost ground: ceding 91 acres to the National Hydro-Electric Power Corporation (NHPC) for its environmentally disastrous damming of the Teesta River; and ceding smaller parcels (5 acres) to an Industrial Training Institute (ITI) in 2016[21] and a government tourist complex in 2017 (6 acres). These land grabs may be small in proportion to the vast tracts under cinchona's domain, but they portend an unsettling future of loss ahead.

Neoliberal capital is not the only force with designs on the cinchona frontier. In 2008, Darjeeling plunged into a second Gorkhaland Agitation (see Middleton, 2015). Gaining control of the cinchonas was a top priority. Per this vision, these lands were an integral part of the territory of Gorkhaland – at once a site of ethnic belonging and a vital resource for moving the region and its people forward. The Gorkhaland Movement failed to deliver a separate state, but the conciliatory government that was created in its wake (known as the Gorkhaland Territorial Administration, GTA) did gain control of the Directorate of Cinchona and Other Medicinal Plants. This created the awkward situation where the cinchona administration is handled by the GTA, but the land itself is dubiously the property of West Bengal. In this arrangement of nested sovereignties, who really controls this frontier and the prospects of its resurrection remains a source of controversy and confusion.

Many who dwell in this space of uncertainty believe cinchona's end to be near. As one former manager surmised, 'The situation is very, very grave. It's almost dead.' Others readily sound the death knell with terminal proclamations like 'the cinchona chapter is now closed' and 'the cinchona plantations are finished!' Seeing no prospect of the industry's resurrection, these imaginations insist that the plantations' resources, energy, and future be directed elsewhere (see Paprocki, this volume). That said, the Cinchona Directorate has refused to give up, championing

Figure 10.4   The Government Chinchona Factory at Mungpoo, 2017.
Photo: Townsend Middleton.

crop diversification as a means to offset the plantations' massive expenditures. Experimental plantings of rubber, citrus, aromatics, ginger, mushrooms and other crops are now underway. The Directorate even ventured a trial-run of the Mungpoo factory in 2014, awakening the sleeping industrial giant from its state of ruination (Figure 10.4).

The trial-run showed that the trees and factory could still produce quinine. However, experts familiar with the science and economics of quinine production told me there was no way that this cinchona frontier could compete with the higher-yield cinchona and cheaper quinine coming from places like the Congo. After many entitlements won, the cost of labour is too high. After decades without technical work, the skill-levels are too low. After life-cycles of agronomic neglect, the bark is too weak. And after years of rot and obsolescence, the factory's infrastructure is not up to pharmaceutical standards. Amid these frontier remains, any talk of cinchona's rebirth was, they insisted, a 'political drama' meant only to keep hope – and the status quo – alive. But to what ends, no one knows.

With their fate undecided, the cinchonas are seemingly stuck in a no-man's land between life and death. Only, this is no no-man's land, but rather an inhabited space, replete with everyday trials and triumphs. That there is life in these remains – human, botanical, material, and otherwise – may perhaps give us pause to rethink the temporalities through which we recognize the lives and ostensible 'deaths' of frontiers. Cinchona suggests that frontier assemblages do not die, so much as fall apart, leaving remains that while diminished of their vitality, nevertheless remain open to re-assemblage, repurposing, and rebirth. In these fathomings of Frontier 2.0, the open-ended, tenuous ontology of a frontier assemblage calls forth its analogous temporality – one attuned not only to the un/makings of a resource frontier, but also its potential remakings.

Recursive Assemblages

Amid these lives and 'deaths', recursivity figures is an analytic with political stakes. There is no question of the imperial lingerings at hand. But for the people of cinchona, the 'debris' of this once-vital medical frontier are the grounds from which the next frontier – or at least the next generation – must grow. It is 'what they have been left with' (Stoler, 2016: 348).

I accordingly offer *recursive assemblages* as an analytic for sorting through the histories, lives, and futures of these frontier remains. Attending to the assemblages and recursivities that produce frontiers at the margin of Asia and beyond allows us to engage not only life amid these remains, but also the life *of* these remains – human, material, and otherwise. If recursive histories are those that fold back upon themselves – as Stoler, has it – *recursive assemblage* helps us appreciate the tenuous, ever-changing, and re-articulating elements in the fold. As cinchona's protean history illustrates, the manifold relationships that constitute frontiers are neither stable nor fixed. New forces are constantly entering the assemblage, disturbing its internal balances, creating new articulations, and changing the nature of the frontier all the while.

Importantly, cinchona's communities remind us that there are more and less desirable forms of life amid these ruins. That they have fended off privatization to defend their lifeways against the tea industry is a case in point. Cinchona may be a plant with a different imperial history than tea, but, thanks to the politics of its communities, cinchona also represents a fundamentally different present – a life worth fighting for, no matter how tenuous the successes (see also Rubinov, this volume).

Amid the ravages of neoliberalism, holding one's own in the face of forces looking to re-capitalize old frontiers may itself constitute success. This is particularly the case at the margins of Asia and beyond, where local communities typically have less recourse to legal protections from the onslaught of privatization and various extraction regimes that prowl the contemporary. What some have deemed the 'failure' of cinchona's trade unions to steer the plantations out from the traps of a bygone frontier may here – precisely by forestalling the 'death' of the frontier – be a victory in and of itself. Those battles have, at the very least, bought some time for another future, another frontier to come into view.

Toward that end, experimental repurposings are now cropping up across the precipitous landscapes where cinchona once dominated. None have yet found the footing to prompt a paradigm shift in how the plantations are run and lived. But before declaring these endeavours futile, it is worth remembering that experimentation was how cinchona first took root in Darjeeling. A century and a half later, the question remains: what life is to be found – and made – on this frontier at India's margins? With various stakeholders vying for answers, the lands and lives of cinchona continue their search for that elusive articulation of elements – that historical click – through which the unmaking of *this* frontier shall become the makings of the next. Chronicling the potentials and politics of Frontier 2.0, we would do well to heed the life in these remains – and stay mindful of both the imperial and more-hopeful possibilities wrought by the recursive histories at hand.

Notes

1 'Darjeeling Hills' connotes the sub-Himalayan landscapes of West Bengal's Darjeeling and Kalimpong districts.
2 Cinchona's early botanical history is well-covered: Markham (1880); Duran-Reynolds (1947); Biswas (1961); Brockway (1979/2002); Headrick (1981); Mukherjee (1996, 1998); Honigsbaum (2002); Rocco (2004); Veale (2010).
3 [Bengal, General, Proc40, March 1862]; Mukherjee (1996): 305. For brevity, I reference archival findings by the file's call number [Government, Ministry, Branch, Proceeding#, Date]. Archives consulted: *The West Bengal State Archives* and the *Indian National Library* (Kolkata); *The National Archives of India* (New Delhi).
4 See first-hand accounts of Cross (1871); Markham (1880); Spruce (1908); also Brockway (1979/2002): 112–117; King (1876); Mukherjee (1996): 311–313.
5 [Bengal, Agricultural, PC9, Proc 29, April 1875].
6 [Bengal, General, Misc, Proc 93, March 1862].
7 [Bengal, General, Misc, Proc 73–77 June 1871].
8 [Bengal, General, Misc, Proc 17–18 May 1870].
9 Anderson's successors could not explain the jump. [Bengal, General Misc, Proc 19–22 December 1869],[Bengal, General, Misc, Proc 43–45 June 1870] The number would later be trimmed, signalling the blurred lines between cinchona and the Forest Department.

10   [Bengal, Revenue, Forests, Proc 5–10 June 1865], [Bengal, Revenue, Forests, Proc 5–10 June 1865].
11   [Bengal, Medical, 13, Proc 33–36 March 1878].
12   [Bengal, Medical, 2 m/4, Proc 11–25 September 1892], [Bengal, Medical, 2 m/3, Proc 6–18 January 1893],[India, Home,Medical, Proc 71–74 September 1893].
13   [India, Revenue and Agriculture, Economic Products, Proc 5–9 August 1897].
14   [India, Revenue and Agriculture, Agriculture, Proc 38–41 July 1917].
15   Ibid.
16   [India, Revenue and Agriculture, Agriculture, Proc 8–33 March 1917].
17   Ibid. On Burma plantations [Revenue, Cinchona, Compiled Proceedings,1921 (Apr–Dec)], [Revenue, Cinchona, Proc B46–51, 53–58, 59–65 July 1927], [India, Education, Agriculture, Proc B381–82 March 1932], [India, DGIMS, Store, File39/44/41S, 1941].
18   [India, DGIMS, Store, Proc, File45/24/42-SI(II),1942].
19   Ibid. [India, DGIMS, Store, File 32/28/39-S, 1939], [India, DGIMS, Store, File 45/33/42-SI(II),1942].
20   Hindustan Lever eventually re-incorporated a small portion of cinchona into their proposal.
21   Envisioning educational and economic benefits, locals generally welcomed this initiative. During construction, however, relations soured and the half-built ITI center was torched in June 2017.

11

Frontier Making and Erasing
Histories of Infrastructure Development in Vietnam

Christian C. Lentz

Introduction

Now a leading source of Vietnam's electrical power, the Black River begins in China and runs 1,000 km before joining the Red River and spilling into an alluvial plain. Draining mountainous area over 50,000 km², it is the Red's largest tributary by volume (Castelleti et al., 2012). Steep topography and heavy flow make the Black an ideal source of hydropower. The first dam at Hòa Bình, begun in 1979 and completed in 1994, accounted for 15% of Vietnam's total electric capacity. The Sơn La Dam, finished in 2012, accounted for 10% national capacity, and stands as Southeast Asia's largest hydroelectric project. The Lai Châu Dam, which came online in December 2016, completed construction on the Black's mainstream. Designed to work in cascade by linking reservoirs upstream to the Chinese border, these three generators are the largest in Vietnam's national power system (Dao, 2011; EVN, 2017).

Rapid dam construction since the 1970s signifies a hydropower frontier but erases the work an earlier resource frontier did, and still does, to make the present one possible. For centuries lasting through the colonial era, diverse highland communities and ruling Tai elites had kept lowland Kinh political projects at a distance. But the distance eroded rapidly in the First Indochina War (1946–1954) when Vietnam's independence struggle against France surged into the Black River region and reached a climax at Điện Biên Phủ (see Figure 11.1).

Frontier Assemblages: The Emergent Politics of Resource Frontiers in Asia,
First Edition. Edited by Jason Cons and Michael Eilenberg.

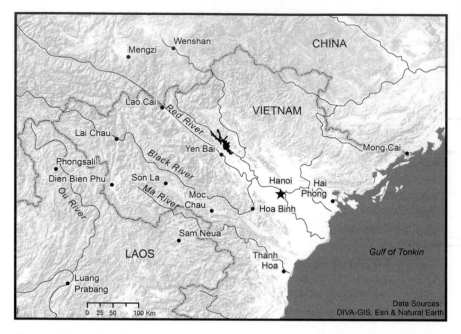

Figure 11.1 The Black River region in Vietnam. *Source*: © Christian C. Lentz. Map by Amanda Henley and Hannah Reckhow.

Fighting near the Lao and Chinese borders and far from secure base areas, Vietnamese forces secured logistical resources locally by integrating the region into its military rear. They connected domestic realms to the military front by mobilizing labour to build transportation infrastructure and appropriating household food for rations. In so doing, tens of thousands of local women and men performed citizenship duties as 'people's labourers' (*dân công*), paid agricultural taxes in kind, and encountered Kinh peoples in positions of command. As they helped roll back colonial rule in everyday ways, the labourers embodied fraught subject relations on an embattled frontier of Vietnamese territory. Their actions during and after war also enabled riverine resource exploitation in ways no one had ever imagined.

The long gap and intervening social changes between these two frontiers raise questions about how to link the 1950s with an unfolding present. How are these two moments related to one another? What is the former's influence on the latter? Where did infrastructure developments take place, and why are their spatial and temporal traces so hard to find?

By considering questions of infrastructure and exploitation on Vietnam's Black River, this chapter analyses the making of a contemporary frontier assemblage and the erasure of its historic antecedent (see also Rubinov and Middleton, this volume). To understand a rapidly-changing frontier requires thinking historically about its production, including warped notions of time and space. As Anna Tsing writes of resource booms in Kalimantan, frontiers are 'notoriously unstable' and advance through a 'series of nonlinear leaps and skirmishes' (2003: 5101–5102). Neither natural nor inevitable but, rather, relational and always in the making, frontiers operate at multiple spatial scales through cyclical and highly-compressed temporalities (Tsing, 2000). As a result, political ecologists have advanced novel ways to conceptualize forces, discourses, actors, and environments as they come together to drive frontier development across Asia – from constellations and conjunctures in Indonesia (Eilenberg, 2014; Li, 2014a) to assemblages in Laos and Burma/Myanmar (Barney, 2009; Woods, 2011b). Here, in keeping with an idea of assemblage, frontiers emerge as dynamic spaces when various parts come together to form a territorial whole (Dovey, 2010: 16–18). Much like regional analogs, then, Vietnam's hydropower frontier is formed from historic and ongoing ecological, political-economic, and cultural processes (see also Swanson and Anderson, this volume). These processes first intersected over half a century ago, yet their traces have become increasingly obscure.

By tracing how a formative history became hidden, I illustrate how erasure not only shapes a frontier assemblage but limits what we as scholars know about its shaping. Itself a process, erasure obscures and silences pasts that continue nonetheless to influence the present (for a more future oriented take on this argument, see Paprocki, this volume). In the mid-twentieth century, coercion and consent drove processes of decolonization, militarization, and nation-state formation in the Black River region, making it Vietnamese (Lentz, 2011a). Without tapping household resources and constructing citizen-subjects, the Democratic Republic of Vietnam (DRV) could not have mustered the logistics to defeat France, secure strategic territory, and achieve independence in 1954 (Lentz, 2011b). As of the early twenty-first century, martial valour is celebrated but contests on the homefront have been silenced – even as the latter continues to etch the landscape anew. On the one hand, officials remember national unity and forget regional dissent by burying documents in the archives, exalting companies in newspapers, and negotiating with Tai powerbrokers (Ha, 2016). On the other hand, local peoples who sacrificed in wars have lost ancestral homes, fisheries, and gardens to rising waters (Dao, 2015a). They endure resettlement to far-away sites often named after places they just evacuated, scrambling maps and destabilizing toponyms. Not simply an accident, erasing a past that explains the present helps legitimate uneven development and secure national territory in tense sites characteristic of what Jason Cons calls 'sensitive space' (2016). Conversely, recovering this past and tracing its links to the present, as I do by analysing newly-available archival records, offers a method to understand frontier assemblages (see also Middleton, this volume).[1]

This chapter excavates pivotal histories and contingent outcomes concealed within but enabling of the Black River's hydropower assemblage. It opens at a conjuncture when, in late 1953, the People's Army stormed the region, toppled a local form of French colonial rule, and mobilized diverse peoples to produce Vietnam's national territory. Focusing on the town of Lai Châu in 1954, the discussion advances an idea of the homefront as a frontier assemblage operating at the scale of households and gendered bodies. It also analyses the unintended outcomes of such foundational work, especially a countermovement led by swidden cultivators from 1955 to 1957. The next section discusses a powerful vision of infrastructure development aired in 1958 and analyses the processes through which its realization has been largely erased. A short conclusion discusses how studying historic erasure in hand with frontier production offers a methodological tool to enrich analytics of conjuncture and assemblage.

Making a Homefront, Generating Vietnamese State Power

The Black River region occupies an ambivalent space in Vietnamese society and national memory. The rugged region is marginal to lowland population centres, hard to reach overland, and culturally different from dominant Kinh norms. By contrast, the region is central to the nation and its history because Vietnam won independence there at the battle of Điện Biên Phủ in 1954. Such marginality and centrality speak to a sense of 'remoteness', understood anthropologically as a discourse of spatial distance underlain by relative familiarity and association (Ardener, 2012; Harms and Hussain, 2014). This related-ness is a geographic function of Vietnamese territory – a spatial, cultural, and political ordering produced in the First Indochina War. By returning to this formative moment when a homefront was constructed at Lai Châu, this section uncovers a place glossed as remote and excluded from national narratives yet – now, as before – a significant source of Vietnamese state power.

December 1953 was a conjuncture in Black River region history. For centuries until then, Tai elites had ruled the region and governed land and labour in valley settlements, known as *muang*. Beginning in the late nineteenth century, French indirect rule preserved the *muang* and incorporated its elites, led by the Đèo clan of Lai Châu, into colonial administration. In 1948, France established the Tai Federation and Đèo Văn Long assumed its presidency. Yet World War II had destabilized the colonial order, enabling Vietnamese activists to join forces with locals alienated by Long, whom they denounced as a traitor. After Long fled into exile in November 1953 and the People's Army took over in December, Lai Châu was no longer capital of a semi-autonomous unit in France's overseas empire. Rather, DRV administrators demoted it to a seat of Mường Lay District – just a humble town in Vietnam's hinterland – but, as Vietnamization of the name indicates, preserved the socio-spatial form of the Tai *muang* (i.e. *mường*). Over the next year, cadres constructed political and transportation infrastructure on this Black River frontier, laying the foundations for larger infrastructure projects still to come.

While the world focused on the battle at Điện Biên Phủ 100 km to the south, Lai Châu endured social and spatial transformations that lay the groundwork for hydropower's emergence there decades later. Having breached the colonial front, the People's Army put Lai Châu's people to work supplying its rear – a site for corralling logistical resources and constructing transportation infrastructure in support of war. In ways similar to Cold War conditions in the United States, constructing a military rear opened a lasting homefront at intimate scales of households, hearths, and bodies (McEnaney, 2000). Now caught inside an emerging and embattled corner of Vietnamese territory, Lai Châu's people became a significant source of labour for the army, their food its rations. Tapping these domestic resources enabled Vietnam's victory over France in May, an epic stride to national independence. Doing so also connected the Black River borderlands, since then known as Northwest Vietnam, to downstream centres of power and population. Indeed, mobilizing the homefront during and after war generated fraught subject relations and built instability, tension, and uncertainty into nation-state territory – all of which remain components of the Black River's frontier assemblage.

Soldiers, cadres, and residents in 1953–1954 Lai Châu lived in a world turned upside down, its order inverted by revolutionary force. Unlike elsewhere in Vietnam and owing to earlier Tai Federation strength, guerilla organizing did not precede 'liberation' by regular troops, an event many locals saw instead as 'occupation'. As a result, cadres assigned by the army to construct Mường Lay District found work 'heavy' among its 20,079 residents. The colonial legacy and ongoing anti-colonial warfare bedevilled the project of national unity. First, ethnic difference mapped onto economic specialization: Tai farmers grew wet-rice in the valleys, Hmong and Dao cultivated swiddens on the mountains, and Chinese traders ran shops in town. Second, because able-bodied men had been drafted by colonial forces, the population was two thirds women, many of whom were unaccustomed to working outside the home. Third, years of war had exhausted the economy, rendering supplies of everyday staples scarce, their prices dear. Fourth, residents had abandoned their homes for the relative safety of forest and field. Owing to communal hatreds stoked by the colonial regime, many feared the worst from revolutionaries. Many Tai residents, for example, were worried that Kinh people would shoot them on sight, then eat their flesh.[2]

Cannibalism notwithstanding, the idea that Vietnamese soldiers were hungry and combative was exaggerated but not entirely untrue. Even after the frontlines had shifted, politics in the rear was violent. The army waged a campaign to 'eliminate bandits', aiming to kill commandos, regime loyalists, and other hostile opponents.[3] This coercive background complemented intensive, front-stage legitimation work among residents. Officers and officials began the latter by 'calling on the people to return to normal lives'. In each village, they organized meetings for the 'confused and suspicious' people to 'study government policy'. Every resident attended, on average, four study sessions.[4] Rather than eat people, residents learned that the DRV sought their bodies for labour and crops for food.

Turning the homefront into a source of appropriable labour and food constructed relations of citizen and subject. Officials tried to persuade the population, first, that the rising state stood for social progress and, second, that it required things in return. They heard how the recently-toppled Đèo family had held the lion's share of good land, learning to associate colonialism with poverty and unjust exploitation. Officials then redistributed land, livestock, and grain to poor peasants, delivering on promises to punish traitors and guarantee equality. Residents also learned about 'government' and its commitment to development, before receiving farm tools. In exchange, cadres explained citizenship duties of *dân công* labour service and agricultural taxation. Cadres derided colonial corvée for pauperizing the poor and, by contrast, praised the 'voluntarism' of service required of all people in fair and equal measure.[5] Working as a *dân công*, then, embodied the rights and duties of national citizenship as well as the subject relations of the postcolonial state. Likewise, whereas colonial taxation had fleeced the peasantry, the DRV's agricultural tax assessed grain levies on a progressive scale.[6] Appropriating household food and labour generated the logistical resources necessary for war against France and, in time to come, stoked contests inside Vietnam over nation-state rule.

Far from returning to ordinary, as propaganda would have it, constructing Vietnam in time of war demonstrated that Lai Châu's state and society were changing in extraordinary ways. For the newly-emerging state, a rapid transition opened gaps between policy and practice. In a December bulletin, Province Chairman Lò Văn Hặc announced 'heavy but honorable duties' required to liberate Lai Châu Province from colonial rule. Whereas people in the province's older free zones had to pay taxes and work as *dân công*, Mường Lay's newly-liberated status supposedly exempted the district.[7] Contrary to this order, however, people in the district, including Lai Châu Town, would in fact work as *dân công* and pay taxes, fuelling popular confusion and anxiety. More broadly, what cadres presented as a singular 'government' was itself composed of competing bureaucratic segments, as illustrated in a case of disputed property. Civilian administrators claimed homes once belonging to rich 'enemies' and colonial collaborators, including those of the now discredited Đèo family. Before they could claim the prize, however, the army sent reinforcements and a hodge-podge of units moved in and used the plush houses to 'rest their feet'. They wound up destroying the houses beyond use. After the units moved out, an administrator expressed his anger through clenched teeth, inviting superiors to 'consider' his military counterparts.[8] His written response registered rivalries and tensions within a Vietnamese state too often glossed as monolithic. As for the event in question, smashing material symbols of colonial domination figures in a long, tense process of erasure that, to this day, obscures an important chapter in Lai Châu's past.

Severed from France, local society became bound ever tighter to Vietnam through mandatory logistics work and the transportation infrastructure they built. Because Lai Châu was a node in borderlands trade and transport networks, district residents bore especially heavy logistics duties. Put to work on road crews and as porters, they did the hard work of moving supplies to the Điện Biên Phủ front. During the war's peak phase, from December to June, over 4,000 residents, or one fifth of the district's population, worked from one to three months. They hefted loads on foot, tended mule trains, built a 90-km road, maintained bridges, and plied rivers on watercraft (see Figure 11.2). All the while, they dodged airplane fire, navigated rapids, and scaled waterfalls. Even after victory in May, *dân công* service continued. First, Mường Lay's residents supplied bandit eradication campaigns until the Geneva Accords ceasefire in July. Then, service continued through the year's end, evolving beyond wartime logistics into peacetime infrastructure work as well. Over 1954, the district registered an average of over three off-farm labourers from each and every household.[9] The human toll was heavy, and still rising.

The state practice of recruiting household labour not only produced a Vietnamese homefront but built tensions deep within the frontier assemblage. 'When the *dân công* hit the road', observed an official, they 'did not have a spirit of mobilization' nor were their anxieties resolved'.[10]

Figure 11.2 Sampans arriving in Lai Châu and discharging rice bound for the Điện Biên Phủ front in 1954. The site is now flooded by dam reservoirs. *Source*: Image courtesy of the National Archives of Vietnam, Center 3.[11]

Indeed, wartime urgency led officials to bypass procedures meant to explain policy and adjust household assessments. Kinh cadres held only a few sessions and mostly in Tai communes, meaning that non-Tai peoples did not have access to outlets for grievance and negotiation. Among highland Hmong and Dao peoples, then, *dân công* service must have felt all too much like the dreaded colonial corvée – a practice many had avoided by substituting tax payments to French officials (Brocheux and Hémery, 2009). In short, the DRV reversed precedent by requiring the highlanders to labour far from home in dangerous and unfamiliar conditions. Many Hmong workers fell ill, and some even died of exhaustion and disease.[12]

In all, widespread reports of popular 'anxiety' indicate an embodied response to labour appropriation as well as the contested legitimacy of state resource claims levied on one's person (MacLean, 2013). Contrary to stated principles, hectic mobilization resulted in many policy 'errors', including unequal service durations and recruiting nominally exempt workers, such as pregnant women. Further, diverting labour from household production damaged agricultural output and threatened social reproduction. 'Going on *dân công* this much', protested one Tai village head, 'will destroy livelihoods'. Some refused to work. Others fled, including six Hmong families who, 'fearing death' by *dân công*, escaped across enemy lines.[13]

The homefront grew still more restless when household resource claims expanded from labour to food. During the campaign, quarter-masters borrowed over 100 tons of rice, vegetables, and meat from Mường Lay's residents to provision soldiers and other *dân công* labourers.[14] The government intended these loans to count towards future taxes, but the unintended consequences came first. That summer, provisioning amplified seasonal hunger among farm households, especially Tai peoples. So, the district scrambled to find food aid by borrowing corn from Hmong growers and buying up rice. Meanwhile, the same problems accompanying *dân công* duties also compromised the autumn tax season. As harvest approached, farmers throughout the province were 'anxious' while those in the district were 'the most scared'. Officials groused about alleged 'enemy counterpropaganda' that 'destroy[ed] tax and *dân công* policies'.[15] But dismissing critiques as traitorous denied valid concerns about hunger and obscured debates about property rights under the new regime. It also overlooked the many ways in which Kinh cadres simply did not understand swidden agriculture and wound up filling tax coffers at the expense of sound relations with Hmong, Dao, and Khmu cultivators. In the end, 1954's agricultural tax yielded tons of grain but the process was riddled with misunderstandings, inequities, and errors.

Not simply acts of extraction, exacting state claims on the homefront drove household transformation and mass political participation in Vietnam's Black River frontier assemblage. Motivated by egalitarian gender ideals and an interest in expanding the pool of eligible labourers, cadres targeted women for *dân công* service. They also sought to impart modes of collective labour and political action. According to a veteran's memoir, 'women outnumbered everyone' on the Black River Front where they built bridges, repaired roads, and served the frontline.[16] Though she acknowledged drudgery and fear, the young woman took pride in her time away from home: 'we were *dân công* not because we had to be but to become the Youth Pioneer Workteam'. In other words, she developed a sense of responsibility outside the family and embraced a purpose beyond the household. Such transformative experiences emboldened women who, under colonial rule, had been subject to gendered violence and racialized hierarchies.

Women's participation in a socialist labour regime elicited as much anxiety as enthusiasm, generating ambivalence about Vietnamese authority. In late 1954, Mường Lay's District Chairman complained how cadres endeavouring to 'exploit women's capacities' felt frustrated that they 'could not yet mobilize women'. Yet his own data tell a more complex story. Though correct that few women joined road crews early on, participation rates were steadily increasing across the board. By December, no less than 300 of 500 workers on construction sites were women. Meanwhile, political education reached more women than men, and women dominated workteam membership.[17] On roadside, classroom, and farm, then, women learned about state priorities of agricultural intensification and collective labour, both of which aimed to increase economic production. Doing so enhanced tax receipts, challenged patriarchal norms, and built infrastructure at bodily scales. Whereas Tai women in Lai Châu had long tended to home and hearth, a police report noted that, after a year of organizing, 'women raked fields, threshed rice, and hefted the burdens of *dân công*.'[18] Yet the additional work they did was not entirely their own.

By the end of 1954, revolutionary warfare and state formation had produced a tense frontier assemblage in Lai Châu, propelling relations with Vietnam on an unstable trajectory. A town that had been a colonial capital only a year before was now a national district, its old Tai powerbroker replaced by a new one, and its people newly liberated and occupied. Working as *dân công* and paying taxes on food facilitated the flow of logistics in war and intensified upstream-downstream linkages long past its end. Moreover, these infrastructural labours enlisted bodies and households in a larger territorial project, constructing a homefront inside Vietnam out of peoples not wholly Vietnamese. Women and men responded ambivalently to nation-state rule, and their ongoing negotiations over the terms of citizenship questioned the relationship between nation and ethnicity as well as rights to and claims on household resources. Even though hydropower was not yet on the state's agenda, the construction of necessary subject relations and infrastructural linkages was well under way.

In this atmosphere of heightened expectation, material scarcity, and political ferment, a DRV decision to initiate discussions on regional autonomy unwittingly kindled a transborder countermovement. Aired in early 1955, the idea and possibility of self-rule sparked debates among ethnicized Black River communities about their place in Vietnam more broadly. Initially, Hmong, Khmu, and Dao communities favoured strong relations with Kinh cadres from Hanoi to fend off entrenched Tai elites flexing *muang*-based powers.[19] But Kinh cadres and DRV leaders were preoccupied with downstream land reform, severe budget deficits, and a shattered economy. As a result, Tai power waxed with the creation of the Thái-Mèo (Tai-Hmong) Autonomous Zone in May 1955. Acting again as mediators in the strategic borderlands, Tai officials implemented central plans and ratcheted claims on agrarian resources – from mobilizing more *dân công* labour for infrastructure to taxing more swidden crops, including opium in 1957 (Lentz, 2017). Somewhere along the way, state claims on the homefront crossed a threshold.

Inspired by radical notions of equality and emancipation yet increasingly burdened by citizenship duties, swidden cultivators embraced their own idea of autonomy by launching a revolutionary alternative to nation-state rule. Referred to as 'Calling for a King' in Vietnamese archival documents, the movement sought self-determination for a political community uniting Hmong, Dao, and Khmu peoples against Tai powerbrokers and Kinh cadres. Echoing highland Southeast Asia's millenarian traditions (Lee, 2015), the swidden cultivators critiqued Tai-Kinh regional domination, protested people's republics in Vietnam and Laos, and appealed for deliverance from a supernatural sovereign.[20] Some followers invoked force, drawing on martial skills learned in anti-colonial warfare. More were peaceful, fasted, prayed, sacrificed livestock, and even let crops go to seed. Refusing to work, they cast off the DRV's productionist agenda, rejected its claims on household resources, and embodied an alternative politics of community self-reliance. Activities escalated until spring 1957 when a violent turn of events prompted a regional crack down (Lentz, 2019). By deploying armed units, policing borders, surveilling swidden communities, and jailing movement leaders, the DRV secured its rear area and reined in the homefront. None of these participants imagined a future featuring dams and resettlement, even as they all made it possible.

Erasing a Frontier

Scholars know little about the mid-1950s Lai Châu homefront and Black River political unrest – for good reason. The dialectical logic of Vietnamese nationalism, and its expression in official history and memory, leaves little room for nuance (Pelley, 2002). Lai Châu is hidden in the shadow of Điện Biên Phủ, even functioning as its silent twin. The latter is celebrated as site of Vietnam's final victory over France and the full liberation of all the region's peoples. As a vanquished capital in French empire once led by the collaborating Đèo clan, the former functions as a reviled site of colonial duplicity, feudal exploitation, and ethnic division. Not just spatial, the binary is also temporal, pivoting on a rupture separating colony from nation, occupation from liberation, and bondage from freedom. After the People's Army hoisted the DRV flag atop the French command's bunker on 7 May 1954, exploitation ended and progress began – or so the story goes. Official memory of these places at this particular moment preserves a myth still central to party-state legitimacy in Vietnam. Any countervailing evidence and narrative has long been suppressed and, as of late, submerged under reservoirs on the Black River.

The erasure of Lai Châu's postcolonial past began when army units ruined Đèo property and continues to this day. Since the destruction of grand homes and looting of luxury furniture in early 1954, the ruination of Đèo lives, property, and reputation in Vietnam is nearly complete, as Philippe Le Failler has documented (2011, 2014). Two of Đèo Văn Long's sons were captured, tried publicly, sentenced to death, and executed immediately. Exiled to France, Long passed away outside Toulouse in 1975. Survived by his daughters, the family's political memory has faded. On the ground in Lai Châu, all that remains of Long's residence is a crumbling stone foundation overgrown with weeds. He is remembered as a 'tyrant' and a 'traitor' – when remembered at all.

Led by the victorious general at Điện Biên Phủ, the processes of historical erasure expanded and accelerated in the wake of regional political unrest. Shortly after the crackdown on Calling for a King, General Võ Nguyên Giáp visited the Black River region and introduced a powerful vision of its future. On 1 June 1958, in a rousing speech before cadres in the Thái-Mèo Zone, he unveiled a 'dream' untethered from the region's past.[21] Hailing the DRV's commitment to socialist development, he envisioned hydropower dams on the Black River, tractors ploughing collective farms, cattle grazing grassy slopes, migration from crowded lowland to sparse upland, improved roads and airfields, miners plundering underground riches, and people cured of malaria. Of all these infrastructural schemes proposing to transform the region's 'backward' economy and speed its integration with Vietnam, hydropower was the most ambitious. The general admitted that achieving the dream would be difficult, but expressed faith in Marxist-Leninist principles and the ability of Tai powerbroker Lò Văn Hặc to steer the region towards a better future. In a speech that must have lasted hours, his only mention of regional history referred to the 1954 military victory, as though that alone triumphed over all before or after it.

If the General's 1958 speech offered an inspiring vision of a socialist future, his speeches in 1959 and 1961 betray a hard-nosed attempt to mould a recalcitrant present in its image. On the great victory's fifth anniversary, President Hô Chí Minh delivered a message of party-state leadership, national unity, and socialist prosperity while the general introduced a program called, euphemistically, 'Democratic Improvement'. Targeting swidden cultivators in particular, the program aimed, economically, to collectivize land and labour; politically, to centralize planning and purge local government; and, militarily, to secure borders and enhance police surveillance.[22] In short, he announced the DRV's intent to consolidate control over resources and governance in Northwest Vietnam – with force as necessary. Two years later, however, in another long speech short on historical understanding, the general admitted his dream was still unfulfilled. Eschewing culpability, he blamed persistent poverty on underperforming cadres and ethnically-divided peoples. Only by unifying as a nation, developing economically, and achieving socialism, he argued, would the dream be realized.[23] Moving to a prosperous and egalitarian future, in other words, meant overlooking social difference, ignoring inequality, and leaving the past behind.

Since the early 1960s, traces of the Black River's early transformation into Vietnam's northwest frontier have become harder to find for reasons of geopolitics, administration, and flooding. First, the DRV fought the breakaway southern Republic of Vietnam, entangling a long and violent civil conflict in the Cold War. Although waged largely in the south, this Second Indochina War (1960–1975) reverberated through the region because of the home front's renewed mobilization and because of fierce combat in Laos, likely involving former participants in Calling for a King. Meanwhile, a slew of campaigns echoed the general's vision and operated again on the homefront, from sponsoring migration of Kinh peoples, collectivizing land and labour of Tai peoples, to prohibiting the swidden cultivation and fixing the settlement of Hmong, Dao, and Khmu peoples (McElwee, 2016). Not long after Vietnam's reunification, Chinese troops invaded in the Third Indochina War (1979–1980). Penetrating as far as 50 km into Vietnamese territory before beating a hasty retreat, they destroyed the built environment, including much of northern Lai Châu Province. Then, as Vietnam's economy stagnated in the 1980s, state planning shifted from socialism to capitalism, initiating another round of transformations on the homefront, including the decollectivization and privatization of land and labour. Through it all, the state itself has remained a site of contest and a source of power for communities contending with rapid change (Lentz, 2014).

Second, territorial administrators have redrawn internal boundaries and replaced toponyms such that locating Lai Châu is now an exercise in geographic sleuthing. Starting with the DRV's demotion of Lai Châu Town to district seat in late 1953, administrators shuffled its governance before detaching its name from the original location and finally fixing it far away. In 1955, the Thái-Mèo Autonomous Zone abolished provinces in favour of counties, and the town fell under Mường Lay County, not district. In 1962, when the Thái-Mèo was renamed the Northwest, Lai Châu became a province again and Lai Châu Town its capital. The arrangement lasted through the dissolution of autonomous zones in 1975 under a unified Socialist Republic of Vietnam. In anticipation of the battle's 50th anniversary in 2003, administrators carved the southern half out of Lai Châu Province to create a new Điện Biên Province. In 2005, they granted the latter Lai Châu Town, changed its name, and demoted it again to a seat named Mường Lay. Meanwhile, in 2004, a 'new' town of Lai Châu was established 50 km away in an entirely different place, known previously as Phong Thổ. In 2013, new Lai Châu became a city and capital of Lai Châu Province.[24] Meanwhile, 'old' Lai Châu began to slip under waters rising behind the gargantuan Sơn La Dam.

The flooding of the Black River valley amounts to the third and final erasure of settlements dating back centuries, literally submerging any material evidence of their historic relations with France, China, Laos, and any other hegemons except Vietnam. As water began rising behind the Hòa Bình Dam in the 1980s and then rose again behind subsequent dams upstream, Giáp's most ambitious vision of infrastructure development was fulfilled, but not entirely as anticipated. Like the water that generates it, electricity flows largely downhill, illuminating cities and cooling homes on the Red River Delta, especially Hanoi. But water now flows upstream as well, especially during the rainy season – displacing residents, flooding fisheries, and inundating once verdant valleys (Hydropower in Vietnam, 2015)[25]. The Hòa Bình Dam has displaced 89,720 people and flooded an area of 750 km²; the Sơn La Dam 92,301 people and 440 km²; the Lai Châu Dam 5,867 and 50 km² – and rising (WLE, 2017). These figures do not include the many smaller but still significant hydropower projects on the Black's tributaries. Displacement has contributed to a renewal of resource contests, political ambivalence, and communal tensions (Dao, 2015b). As landscapes are submerged, so too are the traces of a frontier's violent making and its cyclical remaking.

Conclusion

Known today as Northwest Vietnam, the Black River region now appears like almost any other resource frontier in Asia's borderlands where economic development is booming alongside its costs. Hydropower generates electricity and stimulates industry but floods farmland, ruins fisheries, and displaces settlements. Expanding rubber, coffee, and tea plantations capitalize on market prices but cause conflicts with local communities over property rights. Road construction facilitates mass tourism, migrant access, and commodity exchange, all of which intensify up/downstream linkages for better and worse. All these recent developments grow out of a moment when their completion was not even foretold yet rendered possible nonetheless.

Virtually overnight in December 1953, a frontier assemblage first emerged on the Black River region. When the town of Lai Châu changed from colonial capital to district seat, the transformation augured the rise of DRV regional rule and the production of infrastructure at the scale of homes and bodies. Initiated in war but lasting long beyond, processes of militarization and mobilization as well as embodiment and erasure transformed the historic borderlands into a cutting edge of Vietnamese territory. Embodied struggles over land and labour in the military's rear helped defeat colonial forces at Điện Biên Phủ, produced national citizens out of non-Kinh peoples, and opened a hinterland to state exploitation. Yet everyday resource struggles on the homefront unintentionally spawned a countermovement for self-rule from 1955 to 1957, underscoring the contingency of outcomes (see also Zee, this volume). General Giáp's powerful dream of dam development took 60 years to realize and bore untold costs, underlining the non-linearity of frontier making. Moreover, understanding how these events influence contemporary development and exploitation is, itself, far from straightforward. Nationalist commemoration works in hand with acts of historical erasure to exalt Điện Biên Phủ, silence Lai Châu, and befuddle scholarly efforts to connect their pasts with an unfolding present.

More than just uncover histories lurking below the present's muddied surface, this chapter demonstrates how a particular history has been obscured even as it still shapes a contemporary frontier assemblage. The past is not immediately available to scholars of frontier production, particularly when histories are full of ambiguity, violence, and unintended consequence. By bringing historical erasure to the fore and uncovering what is lost in the process, this chapter offers a way to rethink frontiers in relation to conjuncture and assemblage. 'A conjuncture is dynamic but it is not random. There is path dependence', argues Tania Li in her ethnographic study of land and capitalism in Indonesia (2014a: 16). Her strong statement is largely correct, but holds only as far as scholars can determine what actually happened historically to effect a particular outcome. As such, tracing processes of erasure offers a methodological tool to enrich an analytic of conjuncture. Indeed, revisiting the archives with new questions offers a promising analytical approach to trace contingent possibilities, as Jason Dittmer writes. But using historical inquiry 'to undermine the seeming reality of "path dependence" in the present', as he suggests (2014: 316), overlooks how contemporary assemblages may, in fact, be the product of sub-surface forces still unearthed, deeply submerged, or deliberately erased.

Notes

1 Held in the National Archives of Vietnam, Center 3, Hanoi, records are cited below by file number/record group.
2 Lam Sung, 'Bao cao tong ket cong tac trong 1954', 12 December 1954, 90/ Ủy ban Hành chính Tỉnh Sơn La – Lai Châu [UBHC Sơn-Lai].
3 Ty Công an Lai Châu, 'Bao cao cong tac nam 1954', 25 December 1954, 90/UBHC Sơn-Lai.
4 Lam Sung, 'Bao cao...' 12 December 1954, 90/UBHC Sơn-Lai.
5 Ibid.
6 Lee San. 'Bao cao tông kêt thuên N.N. 1954', 31 December 1954, 102/ UBHC Sơn-Lai.
7 Lò Văn Hặc, 'Hiệu triệu của Ủy ban Kháng chiến Hành chính tỉnh Lai Châu', 22 December 1953, 1258/Phủ Thủ tướng (1945–1954) [PTTg].
8 Lê San, 'Trích yêu: BK v/v nha cua dô dacj dungj cuj o Khu Hành chinh Thi xa bi fa huy', 18 June 1954, 542/PTTg.
9 Lò Văn Hặc, 'Báo cáo tổng kết công tác năm 1954', 5 March 1955, 542/PTTg.
10 Lam Sung, 'Bao cao...' 12 December 1954, 90/UBHC Sơn-Lai.
11 3113/Tài liệu ảnh/Bộ Ngoại giao (1945–1975).
12 Ibid.
13 Ibid.
14 Ibid.
15 Huyện Mường Lay, 'Bao cao Tong ket thue nong nghiep', 25 November 1954, 102/UBHC Sơn-Lai.
16 Hồng Hà. *Chuyện Dân công* (19xx). VN59.01218/National Library of Vietnam.
17 Lam Sung, 'Bao cao...' 12 December 1954, 90/UBHC Sơn-Lai.
18 Ty Công an Lai Châu, 'Bao cao...' 25 December 1954, 90/UBHC Sơn-Lai.
19 Tran Quoc Manh, 'Bao cao tom tat tinh hinh cong tac thang 1 1955', 31 January 1955, 90/UBHC Sơn-Lai.
20 Trần Quốc Mạnh, 'Kết quả thi hành Nghị quyết vùng cao,' 22 November 1957, 2777/Ủy ban Hành chính Khu Tự trị Tây Bắc [UBHC KTTTB].
21 Võ Nguyên Giáp, 'Bai noi chuyen cua pho thu tuong Vo Nguyen Giap truoc can bo trung cap va nghien cuu Tay Bac', 1 June 1958, 144/UBHC KTTTB.
22 Võ Nguyên Giáp, 'Tom tat bai noi chuyen cua đ/c Vo-Nguyen-Giap', 8 May 1959, 182/UBHC KTTTB.
23 Võ Nguyên Giáp, 'Bai noi chuyen cua đ/c Vo Nguyen Giap truoc can bo lanh đao cac nganh đang, chinh, quan, dan', 1961, 271/UBHC KTTTB.
24 Anonymous, 'Lai Châu'. *https://vi.wikipedia.org/wiki/Lai_Châu*. Accessed 17 March 2017.
25 'Hydropower in Vietnam: Full to Bursting'. *The Economist* (20 June 2015).

Conclusion

Assembling the Frontier

Michael Eilenberg and Jason Cons

The frontier is always already empty. – Anna Tsing

Rethinking imperial formations as polities of dislocation and deferral which cut through the nation-state by delimiting interior frontiers as well as exterior ones is one step in reordering our attention. – Ann Stoler

This volume responds to the contemporary reimagination of margins and state edges as resource-rich, unexploited 'wastelands' targeted for development schemes for economic integration and control in and beyond Asia. As contributions to this volume attest, these spaces constitute particular kinds of edges where, to follow Anna Tsing, the expansive natures of extraction and production come into their own together with the apparently contradictory logics of containment and securitization (Tsing, 2005). These are zones where sedimented histories of marginality (Moore, 2005), relations of distance and remoteness, and diverse forms of materiality come together with new politics and techniques of sovereignty and capital.

The making of Bangladesh's climate frontier through anticipatory ruination, the grafting of new livelihoods on a former Soviet frontier in Tajikistan, the histories of frontier erasure in Vietnam, and the many other cases in this volume illustrate the diversity of past and present frontier making in Asia and the enormous effects that these processes have on people and environment. These chapters show that resource frontiers are things in the process of becoming: assembled by a diverse

Frontier Assemblages: The Emergent Politics of Resource Frontiers in Asia,
First Edition. Edited by Jason Cons and Michael Eilenberg.
© 2019 John Wiley & Sons Ltd. Published 2019 by John Wiley & Sons Ltd.

array of human and non-human actors, organic and inorganic sub-stances, technical and natural materials, and intangible elements. Moreover, they illustrate the value of approaching frontiers as assem-blages: spaces where the complex causes and consequences of territorial transformation must be charted and demonstrated, not assumed.

Frontier assemblages in Asia and beyond are wildly heterogenous in form, purpose, and shape. Yet, the case studies in this volume show that despite different political regimes (neoliberal/capitalist, socialist, post-socialist) and their attendant economic infrastructures there are striking similarities in the ways resources are assembled for extraction and land assembled for management in these sensitive spaces. In our attempt to uncover similarities across different Asian frontiers, several themes, or drivers of frontierization, tie together the various case studies across the region. For example, several chapters show how the discourse of climate and environmental change have become an explicit or implicit force and rationality for new rounds of assembling resource frontiers. In the case of Bangladesh, Kasia Paprocki shows how coastal landscapes are reshaped into climate frontiers through development planning and notions of resilience and adaptation. In a similar manner, Zachery Anderson charts the Indonesian government's attempt to develop a 'green economy' in its outer islands to mitigate climate change by incor-porating programs of sustainable agriculture and carbon sequestration projects. These projects, driven by strong frontier imaginaries, will open up the region for a new round of resource extraction through biofuel production. Carbon mitigation and climate change also appear as central themes in the chapter by Gökçe Günel, who shows how the United Arab Emirates explores the subsurface of old oil wells as a new imaginative frontier for carbon storage. Similarly to Anderson and Paprocki, Günel's exploration shows that imaginations of the climate affected future are themselves strong forces of frontier making, allowing the conceptualiza-tion of certain spaces as open for frontierization as a hedge against and/ or an experimental site for managing future catastrophe.

Jumping to the arid and rolling dunes in Inner Mongolia, Jerry Zee emphasizes how the climate/environmental crisis of dust affecting major Chinese metropoles are triggering frontier making, or remaking, responses from Beijing through major programs of environmental governance. These programs rework the dusty frontier regions into forest landscapes for future resource extraction and social, economic, ecological, and technical experimentation. This kind of landscape governance, which applies the discourses of environmental sustainability in the making of new frontiers, is further in evidence along China's coasts. Here, as Young Rae Choi shows, large-scale land reclamation projects are reframed as ecologically sustainable practices, despite previous assessments to the contrary. Choi argues that these transformations of oceans into land constitute a form of frontierization – making coastal reclamations invest-able and legitimate test sites for the 'eco' agenda of the Chinese state. Choi and Zee thus both chart processes that articulate with Paprocki, Anderson, and Günel's mapping of climate frontiers. Yet the concerns that drive these transformations are not future catastrophe, but present-day anxieties about atmosphere, population, and investment.

Choi's chapter brings forward another overlapping theme that runs through the book: that of massive infrastructure construction and its role in creating resource frontiers and opening up marginal lands for various forms of extraction and control. Here, the use of infrastructure serves as a means of at once making and settling frontier space. Duncan McDuie-Ra similarly shows how the 'disturbed' and marginal frontier city of Imphal in northeast India is being 'civilized' by the central state through the construction of economic infrastructure and investments in a frontier health industry – an attempt to reclaim the unruly frontier back into the fold of the Indian state. However, as argued by McDuie-Ra, such civilizing processes produce unanticipated effects, which para-doxically reorient Imphal even further towards Southeast Asia and outside the imagination of a sovereign 'India' thus underlining the open-endedness of frontier assemblages. Further, in his chapter on massive-scale mining and urban megaprojects in western China, Max D. Woodworth depicts how spatial transformation and infrastructure pro-jects in frontier space often take 'gigantic' forms as spectacular symbols of development and accumulation. Each of these chapters signals the centrality of infrastructure in frontier space, but also foregrounds a point made in much of the new literature on infrastructure: that infrastructure projects are situated, situational, and open to multiple valences, affects, and interpretations (see Von Schnitzler, 2016; Anand, 2017).

Yet another recurring theme that runs through these chapters is that of frontier temporalities – the binding or disjunctures of past and present frontiers. A vivid example of such temporal alignments of past and pre-sent frontiers is offered by Townsend Middleton in his analysis of quinine production in northeast India under British colonial rule. Middleton traces how this medical resource frontier collapsed and was later reborn and reassembled on the debris of colonial projects. This reassembly is driven at once by a refusal to give up on certain benefits of the older frontier assemblage and by new imaginaries of untapped resources and wealth that speak to the new political and economic realities. In his anal-ysis of frontier making and unmaking in the Pamir mountains of Tajikistan, Igor Rubinov similarly shows how local residents rework frontier ruins into new assemblages after the collapse of the Soviet state and subsequent abandonment of its large-scale infrastructure development projects on the Pamir resource frontier. Histories of frontier making and ruination are also a recurrent theme in Heather Anne Swanson's chapter on the Japanese salmon frontier on the island of Hokkaido. Here, she shows the intricate entanglements of politics, economy and biology in frontier making. Moreover, she illustrates how the Japanese government attempts to breathe life into the ruins of past frontier projects and activate a new round of resource extraction on the salmon frontier. Like the previous chapters, Christian Lentz engages the temporalities of frontier making. As he points out, the rubble of old frontiers may be central in the construction of new frontier assemblages. But it also may be actively occluded, buried, and swept under the rug of nationalist narratives of territory. Through the case of Vietnam's Black River region, he highlights how such frontier erasures became important instruments in Vietnamese state building efforts and infrastructure development/modernization. In all of these cases, frontier assemblies are shown to be contingent and conjunctural. But the things that they assemble have strange lives that live on. Sometimes, these are legacies that are actively engaged in new frontier and post-frontier projects. Sometimes they are ghosts that haunt new frontiers even as they provide the grounds for their making.

Beyond these resonances, contributions to this volume show the value of thinking of frontiers not as self-evident 'things' but rather as assemblages. Resource frontiers are emergent, dynamic, and conjunctural phenomena that fluctuate with the vagaries of politics, markets, and ecologies across time and scale. Through a shared analytic of frontier assemblages, we have traced these emergences – charting the framings and imaginations of space and territory and the historical contingencies that facilitate the transformation of margins into frontiers. Frontier assemblages encompass the specific ecologies of resource frontiers, the forms of capital that underwrite extraction, the specific practices of resource exploitation, and the materiality and ruination of projects unfolding within them. As Li reminds us 'assemblage links directly to a practice, to assemble' and thus indicates the agencies involved in the shaping of connections holding 'heterogeneous elements' together (Li, 2007a:264).

We see frontier assemblages as an analytic that provides different ways to understand frontier dynamics in the contemporary moment. In pairing these terms, we invoke a specific, if open-ended, understanding that frames both a theoretical and methodological approach to understanding resource frontiers. These are sites that at once dramatize and clarify the asymmetries of power and the consequent inequalities and exclusions around resource extraction and production. Our interest in thinking the frontier in this way stems from our collective observation of contemporary territorial transformation across Asia. Yet, we suggest, the kinds of assemblages charted in this book resonate beyond Asia as well. As the explosion of literature on resource frontiers across the globe suggests, we are in a new moment of frontier making. Many of those transformations are framed openly through the language of frontiers. Others mobilize strategies and tactics that map to what we here call fronterization. These emergent frontiers are but the current iteration of a longer historical process: the incorporation of marginal spaces under rubrics of capital, security, and territorial rule. By broadening our understanding of resource frontiers, we suggest that these spaces hold the key to understanding this critical territorial shift.

The dynamics of fronterization are intimately linked with imaginations of these spaces as remote. In other words, resource frontiers are zones in which distance, strangeness, and edginess work to open up conditions of possibility. These conditions are sedimented in the very landscape – built on the ruins of long historical projects of anxious rule. As argued by Tsing, these spaces are relationally produced as 'out of the way' through the dynamics of centre and periphery, legibility and illegibility, and law and lawlessness (Tsing, 1993). In other words, the remoteness of frontiers is socially produced alongside of and through tensions and anxieties about these spaces and the recursive interventions, and their regular failures, that such anxieties herald (Dunn and Cons, 2014). Ardener notes that remote areas are full of the ruins of intervention. 'Remote areas cry out for development, but they are continuous victims of visions of development ... Remote areas offer images of unbridled pessimism or utopian optimism, of change and decay, in their memorials' (Ardener, 2012: 529). In other words, remoteness is an invitation to projects of incorporation, development, integration, and possibility. At the same time, it is a marker of the failure of such projects. Resource frontiers are heterotopic in Michel Foucault's sense of the term: representations of utopian possibility and reflections of an optimistic future. These 'other spaces' are set apart from the rest of society as zones in which different relations of power, and hence different forms of government rationality, can be imagined and implemented. As contributors to this volume show, resource frontiers are often idealized visions of modernity and laboratories of new social ecological orders (Foucault, 1986). At the same time, they are markers of the ruins of progress. Like Walter Benjamin's Angel of History, their histories often read as 'one single catastrophe which keeps piling up wreckage upon wreckage and hurls it in front of [their] feet' (Benjamin, 1968: 257). Yet, like all ruined frontiers, they present themselves as open for new reappropriations and frontierizations – the markers of ruination offer new possibilities and transformative openings.

Beyond progress and ruination, the notion of remoteness is critical to the dynamics of frontier assemblages because it further facilitates a range of intrusions that rely on the trope of distance and remove. What happens at the edges is specifically outside of the bounds of normal intervention. Yet this is not necessarily exceptionality in Agamben's sense of the term (1998). Rather, remoteness facilitates a framing of frontiers as abnormal – spaces imagined as at a remove from state power in ways that necessitate and legitimate a range of actions that skirt the boundaries of law and social norms. In other words, the relations of remoteness are predicated not on an active sovereign decision, but rather on its impossibility – a recognition that state power only partially penetrates and manages these areas or that the interests of the 'state' are only one of a range of competing projects within frontier space. Frontiers often present outsiders with an 'institutional vacuum' of possibilities (Kopytoff, 1999: 33). As Derek Hall notes, the capability to govern frontier spaces is shared among multiple actors, both state and non-state, resulting in an internally fragmented approach to governance that highlight the anxieties surrounding the extraction of valuable resources. States can seldom persuasively claim to be the sole source of law and government in frontier spaces, and projects seeking to incorporate these spaces into sovereign and territorial folds must be resigned to sharing authority with other actors (Hall, 2013: 52). As Danilo Geiger reminds us in many of these marginal spaces, 'colonial regimes never established full administrative control, and thus left their post-colonial successor governments a legacy of still open frontiers' (Geiger, 2008: 93). Frontiers are thus zones characterized by competing normative orders and modes of regulation, producing spaces for resistance, dissonance and manoeuvring (Barney, 2009: 152).

In sum, understanding resource frontiers as frontier assemblages opens up a mode of engagement that highlights the dynamics of frontierization: the ways that frontier spaces are framed and made into sites and zones of production and extraction. We believe that there is significant analytical advantage in thinking the political, economic, social, and material complexity of resource frontiers from this vantage point. Frontier assemblage moves beyond a narrow political economy and instead foregrounds the often surprising collisions of history, ecology, economy, politics, geography, and imagination that come together in frontier space. Attending to these dynamics opens up new analytic, and perhaps political possibilities in emergent frontiers in Bangladesh, China, India, Indonesia, Japan, Tajikistan, the United Arab Emirates, Vietnam, and beyond.

Bibliography

Abrams, P. (1988). Notes on the difficulty of studying the state. *Journal of Historical Sociology*, 1(1): 58–89.

Acheson, J. (1975). The lobster fiefs: Economic and ecological effects of territoriality in the Maine lobster industry. *Human Ecology*, 3(3): 183–207.

Adams, V., Murphy, M. and Clarke, A. (2009). Anticipation: Technoscience, life, affect, temporality. *Subjectivity*, 28: 246–265.

Adnan, S. (2013). Land grabs and primitive accumulation in deltaic Bangladesh: Interactions between neoliberal globalization, state interventions, power relations and peasant resistance. *Journal of Peasant Studies*, 40(1): 87–128.

Agamben, G. (1998). *Homo sacer: Sovereign power and bare life*. Stanford, CA: Stanford University Press.

Agrawal, A. (2005). Environmentality: Community, intimate government, and the making of environmental subjects in Kumaon, India. *Current Anthropology*, 46(2): 161–190.

Ahmed, N. and Troell, M. (2010). Fishing for prawn larvae in Bangladesh: An important coastal livelihood causing negative effects on the environment. *Ambio*, 39(1): 20–29.

Althusser, L., *et al.* (2015). *Reading Capital: The complete edition*. New York: Verso.

Amnesty International. (2013). *The Armed Forces Special Powers Act: Time for a renewed debate in India on human rights and national security*. Bangalore: Amnesty International ASA 20/042/2013.

Anagnost, A. (1997). *National past-times: Narrative, representation, and power in modern China*. Durham, NC: Duke University Press.

Anand, N. (2017). *Hydraulic city: Water and infrastructures of citizenship in Mumbai*. Durham, NC: Duke University Press.

Frontier Assemblages: The Emergent Politics of Resource Frontiers in Asia,
First Edition. Edited by Jason Cons and Michael Eilenberg.
© 2019 John Wiley & Sons Ltd. Published 2019 by John Wiley & Sons Ltd.

Anderson, B. (2010). Preemption, precaution, preparedness: Anticipatory action and future geographies. *Progress in Human Geography*, 34(6): 777–798.

Anderson, B., *et al.* (2012). On assemblages and geography. *Dialogues in Human Geography*, 2(2): 171–189.

Anderson, J. (2009). Transient convergence and relational sensibility: Beyond the modern constitution of nature. *Emotion, Space and Society*, 2(2): 120–127.

Anderson, R., *et al.* (2016). Green growth rhetoric versus reality: Insights from Indonesia, *Global Environmental Change*, 38: 30–40.

Anderson, V. (2004). *Creatures of empire*. Oxford: Oxford University Press.

Araghi, F. (2009). The invisible hand and the visible foot: Peasants, dispossession and globalization. In: Akram-Lodhi, A. and Kay, C. eds., *Peasants and globalization: Political economy, rural transformation and the agrarian question*. London: Routledge. pp. 111–147.

Araki, H., *et al.* (2008). Fitness of hatchery-reared salmonids in the wild. *Evolutionary Applications*, 1: 342–355.

Ardener, E. (2012). 'Remote areas': Some theoretical considerations. *HAU: Journal of Ethnographic Theory*, 2(1): 519–533.

Arendt, H. (1979 [1948]). *The origins of totalitarianism*. New York: Harvest.

Arnold, D. (2006). *The tropics and the traveling gaze: India, landscape and science, 1800–1856*. Seattle: University of Washington Press.

Arnold, D. (2008). *The fishermen's frontier: People and salmon in Southeast Alaska*. Seattle: University of Washington Press.

Arnold, D. (2012). Spatial practices and border SEZs in Mekong Southeast Asia. *Geography Compass*, 6(12): 740–751.

Astuti, R. and McGregor, A. (2015). Governing carbon, transforming forest politics: A case study of Indonesia" REDD+ Task Force. *Asia Pacific Viewpoint*, 56(1): 21–36.

Bach, J. (2011). Modernity and the urban imagination in economic zones. *Theory, Culture & Society*, 28(5): 98–122.

Bailey, I. and Caprotti, F. (2014). The green economy: Functional domains and theoretical directions of enquiry. *Environment and Planning A*, 46(8): 1797–1813.

Baird, I. (2014). The global land grab meta-narrative, Asian money laundering and elite capture: Reconsidering the Cambodian context. *Geopolitics*, 19(2): 431–453.

Barbora, S. (2017). Riding the rhino: Conservation, conflicts, and militarization of Kaziranga National Park in Assam. *Antipode*, 49(5): 1145–1163.

Barney, K. (2009). Laos and the making of a 'relational' resource frontier. *Geographical Journal*, 175(2): 146–159.

Baruah, S. (2007). *Durable disorder: Understanding the politics of Northeast India*. Delhi: Oxford University Press.

Batten, B. (2003). *To the ends of Japan: Premodern frontiers, boundaries, and interactions*. Honolulu: University of Hawaii Press.

Baud, M. and van Schendel, W. (1997). Toward a comparative history of borderlands. *Journal of World History*, 8(2): 211–242.

Bayly, C. (2000). *Empire and information: Intelligence gathering and social communication in India, 1780–1870.* Cambridge: Cambridge University Press.

Bellot, F. (2015). *Drivers of deforestation: Global, national, and local scale,* Jakarta: GIZ FORCLIME.

Belton, B. (2016). Shrimp, prawn and the political economy of social wellbeing in rural Bangladesh. *Journal of Rural Studies,* 45: 230–2242.

Benjamin, W. (1968). *Illuminations: Essays and reflections.* New York: Schocken.

Benjamin, W. (1999). *The arcades project.* Cambridge, MA: Harvard University Press.

Bennet, J. (2010). *Vibrant matter: A political ecology of things.* Durham, NC: Duke University Press.

Bennike, R. (2017). Frontier commodification: Governing land, labour and leisure in Darjeeling, India. *South Asia: Journal of South Asian Studies,* 40(2): 256–271.

Besky, S. (2013). *The Darjeeling distinction: Labor and justice on fair-trade.* Berkeley: University of California Press.

Biehl, J. and Locke, P. (2010). Deleuze and the anthropology of becoming. *Current Anthropology,* 51(3): 317–351.

Billé, F. (2014). Territorial phantom pains (and other cartographic anxieties). *Environment and Planning D: Society and Space,* 32(1): 163–178.

Billé, F. (2016). On China's cartographic embrace: A view from Its northern rim. *Cross-Currents,* 21: 88–110.

Biswas, J., ed. (1961). Cinchona cultivation in India – its past, present, and future. *Journal of the Asiatic Society,* 3: 63–80.

Bliss, F. (2010). *Social and economic change in the Pamirs (Gorno-Badakhshan, Tajikistan).* London: Routledge.

Bora, P. (2010). Between the human, the citizen and the tribal: Reading feminist politics in India's northeast. *International Feminist Journal of Politics,* 12(3–4): 341–360.

Borras, S. and Franco, J. (2011). *Political dynamics of land-grabbing in Southeast Asia: Understanding Europe's role.* Amsterdam: The Transnational Institute.

Botanical Institute. (1948). *Instructions for forest development in Tajik SSR [Instrukziya Po Lesorazvedeniju v Tadjikskoj SSR].* Stalinabad: Botanical Institute of the Tajik National Academy of Sciences.

Bowker, G. (1994). *Science on the run: Information management and industrial geophysics at Schlumberger, 1920–1940.* Cambridge: Cambridge University Press.

Boyd, W. (2001). Making meat: Science, technology, and American poultry production. *Technology and Culture,* 42(4): 631–664.

Bradbury, J. (1979). Towards an alternative theory of resource-based town development in Canada. *Economic Geography,* 55(2): 147–166.

Brand, U. (2012). Green economy – the next oxymoron? No lessons learned from failures of implementing sustainable development. *GAIA-Ecological Perspectives for Science and Society,* 21(1): 28–32.

Braun, B. (2002). *The intemperate rainforest: Nature, culture, and power on Canada's west coast*. Minneapolis: University of Minnesota Press.

Braun, B. (2005). Environmental issues: Writing a more-than-human urban Geography. *Progress in Human Geography*, 29(5): 635–650.

Braun, B. (2006). Environmental issues: Global natures in the space of assemblage. *Progress in Human Geography*, 30(5): 644–654.

Bridge, G. (2001). Resource triumphalism: Postindustrial narratives of primary commodity production. *Environment and Planning*, 33(12): 2149–2173.

Brocheux, P. and Hémery, D. (2009). *Indochina: An ambiguous colonization*. Berkeley: University of California Press.

Brockway, L. (1979/2002). *Science and colonial expansion: The role of the British Royal Botanic Gardens*. New Haven, CT: Yale University Press.

Brown, W. (2010). *Walled states, waning sovereignty*. New York: Zone Books.

Bryant, R. (1997). *The political ecology of forestry in Burma, 1824–1924*. Honolulu: University of Hawaii Press.

Bryson, M. (2016). *Goddess on the frontier: Religion, ethnicity, and gender in Southwest China*. Stanford, CA: Stanford University Press.

Buchanan J., Kramer, T. and Woods, K. (2013). *Developing disparity – regional investment in Burma's borderlands*. Amsterdam: The Transnational Institute.

Burnett, D. G. (2001). *Masters of all they surveyed: Exploration, geography, and a British el dorado*. Chicago, IL: University of Chicago Press.

Büscher, B. (2013). *Transforming the frontier: Peace parks and the politics of neoliberal conservation in Southern Africa*. Durham, NC and London: Duke University Press.

Callon, M. (1986). Some elements of a sociology of translation: Domestication of the scallops and the fishermen of St. Brieuc Bay. *The Sociological Review*, 32: 196–223.

Caprotti, F. (2016). *Eco-cities and the transition to low carbon economies*. New York: Springer.

Carse, A. (2014). *Beyond the big ditch: Politics, ecology, and infrastructure at the Panama Canal*. Cambridge, MA: MIT Press.

Cartier, C. (2001). 'Zone fever', the arable land debate, and real estate speculation: China's evolving land use regime and its geographical contradictions. *Journal of Contemporary China*, 10(28): 445–469.

Casson, A., Muliastra, Y. and Obidzinski, K. (2015). *Land-based investment and green development in Indonesia: Lessons from Berau district, East Kalimantan*. Center for International Forestry Research, Working Paper 180.

Castelleti, A., *et al*. (2012). Assessing water reservoirs management and development in Northern Vietnam. *Hydrology and Earth System Sciences*, 16(1): 189–199.

Census of India. (2011). Census info India. [online]. http://censusindia.gov.in/2011census/censusinfodashboard/index.html

Chang, I.-C. C. and Sheppard, E. (2013). China's eco-cities as variegated urban sustainability: Dongtan eco-city and Chongming eco-island. *Journal of Urban Technology*, 20(1): 57–75.

Chapman, R., Plummer, P. and Tonts, M. (2015). The resource boom and socio-economic well-being in Australian resource towns: A temporal and spatial analysis. *Urban Geography*, 36(5): 629–653.

Chatterjee, I. (2013). *Forgotten friends: Monks, marriages, and memories of Northeast India*. Delhi: Oxford University Press.

Chen, J. (2007). Rapid urbanization in China: A real challenge to soil protection and food security. *Catena*, 69(1): 1–15.

Chen, J. C. (2013). Sustainable territories: Rural dispossession, land enclosures and the construction of environmental resources in China. *Human Geography*, 6(1), 102–118.

Chen, J. C., Zinda, J. A. and Yeh, E. T. (2017). Themed issue: Recasting the rural: State, society and environment in contemporary China. *Geoforum*, 78: 83–152.

Chien, S. (2013). Chinese eco-cities: A perspective of land-speculation-oriented local entrepreneurialism. *China Information*, 27(2): 173–196.

Cho, M. (2006). *Developmental politics and green progressives*. Seoul: Environment & Life.

Choi, Y. (2014). Modernization, development and underdevelopment: Reclamation of Korean tidal flats, 1950s–2000s. *Ocean and Coastal Management*, 102 (Part B): 426–436.

Choy, T. (2011). Ecologies of Comparison: An Ethnography of Endengerment in Hong Kong. Durham, NC: Duke University Press.

Cliff, T. (2013). Peripheral urbanism. Making history on China's northwest frontier. *China Perspectives*, 3: 13–23.

Collier, S. (2011). *Post-Soviet social: Neoliberalism, social modernity, biopolitics*. Princeton, NJ: Princeton University Press.

Colombi, B. and Brooks, J. (2012). *Keystone nations: Indigenous peoples and Salmon across the North Pacific*. Santa Fe: SAR Press.

Cons, J. (2016). *Sensitive space: Fragmented territory at the India-Bangladesh border*. Seattle: University of Washington Press.

Cons, J. (2018). Staging climate security: Resilience and heterodystopia in the Bangladesh borderlands. *Cultural Anthropology*, 33(2): 266–294.

Cons, J. and Sanyal, R. (2013). Geographies at the margins: Borders in South Asia – an introduction. *Political Geography*, 35: 5–13.

Cordell, J. (1989). *A sea of small boats, cultural survival*. Cambridge: Cultural Survival Report.

Cosgrove, D. and Daniels, S. (1988). *The iconography of landscape: Essays on the symbolic representation, design and use of past environments*. Cambridge: Cambridge University Press.

Cramb, R. and McCarthy, J. F., eds. (2016). *The oil palm complex: Smallholders, agribusiness and the state in Indonesia and Malaysia*. Singapore: NUS Press.

Cronon, W. (1983). *Changes in the land: Indians, colonists, and the ecology of New England*. New York: Hill and Wang.

Cronon, W. (1991). *Nature's metropolis: Chicago and the great west*. New York: W. W. Norton.

Cronon, W. (1996). The trouble with wilderness; or, getting back to the wrong nature. In: Cronon, W. ed., *Uncommon ground: rethinking the human place in nature*. New York: W. W. Norton & Co, pp. 69–90.

Crosby, A. (1986). *Ecological imperialism: The biological expansion of Europe, 900–1900*. Cambridge: Cambridge University Press.

Cross, R. (1871). *Collecting of seeds and plants of the Chinconas of Pitayo*. London: Eyre and Spottiswoode.

Cross, J. (2014). *Dream zones: Anticipating capitalism and development in India*. London: Pluto Press.

Crossley, P., Siu, H. and Sutton, D. (2006). *Empire at the margins: Culture, ethnicity, and frontier in early modern China*. Berkeley: University of California Press.

Curzon, G. (1907). *Frontiers: The Romanes lectures*. Oxford: Clarendon Press.

Danermark, B., *et al.* (2002). *Explaining society: Critical realism in the social sciences*. London: Routledge.

Dao, N. (2011). Damming rivers in Vietnam: A lesson learned in the Tây Bắc Region. *Journal of Vietnamese Studies*, 6(2): 106–140.

Dao, N. (2015a). Political responses to dam-induced resettlement in Northern Uplands Vietnam. *Journal of Agrarian Change*, 16(2): 291–317.

Dao, N. (2015b). Rubber plantations in the Northwest: Rethinking the concept of land grabs in Vietnam. *Journal of Peasant Studies*, 42(2): 347–369.

Das, V. and Poole, D. (2004). The state and its margins: Comparative Ethnographies. In: Das, V. and Poole, D. eds., *Anthropology in the margins of the state*. Delhi: Oxford University Press, pp. 3–33.

Datta, A. (1998). *Land and labour relations in South-West Bangladesh: Resources, power and conflict*. New York: St. Martin's Press.

Datta, A. (2006). Who benefits and at what cost? Expanded shrimp culture in Bangladesh. In: Rahman, A., *et al.* eds., *Shrimp farming and industry: Sustainability, trade and livelihoods*. Dhaka: University Press Ltd, pp. 507–528.

De Angelis, M. (2004). Separating the doing and the deed: Capital and the continuous character of enclosures. *Historical Materialism*, 12(2): 57–87.

De Koninck, R. (2006). On the geopolitics of land colonization: Order and disorder on the frontiers of Vietnam and Indonesia. *Moussons*, 9(10): 33–59.

De Koninck, R., Bernard, S. and Bissonnette, J. (2011). *Borneo transformed: Agricultural expansion on the Southeast Asian frontier*. Singapore: NUS Press.

DeLanda, M. (2006). *A new philosophy of society: Assemblage theory and social complexity*. London: Continuum.

Deleuze, G. (2007). *Regimes of madness: Texts and interviews 1975–1995*. New York: Semiotext(e).

Deleuze, G. and Guattari, F. (1986). *Kafka: Towards a minor literature*. Minneapolis: University of Minnesota Press.

Deleuze, G. and Guattari, F. (1987). *A thousand plateaus: Capitalism and schizophrenia*. Minneapolis: University of Minnesota Press.

Development Research Center of the State Council and World Bank. (2014). *Urban China: Toward efficient, inclusive, and sustainable urbanization*. Washington, DC: World Bank Publications.

Dittmer, J. (2014). Geopolitical assemblages and complexity. *Progress in Human Geography*, 38(3): 385–401.

Dobhal, H. (2009). *Manipur in the shadow of AFSPA: Independent people's tribunal report on human rights violations in Manipur*. Imphal: Socio Legal Information Centre.

Donnan, H. and Wilson, T. (1994). *Border approaches: Anthropological perspectives on frontiers*. Lanham, MD: University Press of America.

Dovey, K. (2010). *Becoming places: Urbanism/architecture/identity/power*. London: Routledge.

Duara, P. (2016). Opening remarks, presented at the *Inter-Asian Connections V: Seoul* Conference. Seoul, Korea, April 27, 2016.

Dunn, E. C. and Cons, J. (2014). Aleatory sovereignty and the rule of sensitive spaces. *Antipode*, 46(1): 92–109.

Duran-Reynals, M. (1947). *The fever bark tree: The pageant of quinine*. London: W. H. Allen.

Durrenberger, E. and Pálsson, G. (1987). Ownership at sea: Fishing territories and access to sea resources. *American Ethnologist*, 14(3): 508–522.

Dwyer, M. (2013). Building the politics machine: Tools for 'resolving' the global land grab. *Development and Change*, 44(2): 309–333.

Ehrenberg, S. N. and Nadeau, P. H. (2005). Sandstone vs. carbonate petroleum reservoirs: A global perspective on porosity-depth and porosity-permeability relationships. *AAPG Bulletin*, 89(4): 435–445.

Eilenberg, M. (2011). Straddling the border: A marginal history of guerrilla warfare and 'counter-insurgency' in the Indonesian borderlands. *Modern Asian Studies*, 45(6): 1423–1463.

Eilenberg, M. (2012). *At the edges of states: Dynamics of state formation in the Indonesian borderlands*. Leiden: KITLV Press.

Eilenberg, M. (2014). Frontier constellations: Agrarian expansion and sovereignty on the Indonesian-Malaysian border. *Journal of Peasant Studies*, 41(2): 157–182.

Eilenberg, M. (2015). Shades of green and REDD: Local and global contestations over the value of forest versus plantation development on the Indonesian forest frontier. *Asia Pacific Viewpoint*, 56(1): 48–61.

Elden, S. (2010). Land, terrain, territory. *Progress in Human Geography*, 34(6): 799–817.

Elden, S. (2013). *The birth of territory*. Chicago, IL: University of Chicago Press.

Erman, E. (2008). Rethinking legal and illegal economy: A case study of tin mining in Bangka Island." Unpublished Manuscript. http://globetrotter.berkeley.edu/GreenGovernance/papers/Erman2007.pdf

EVN (Electricity Viet Nam). (2017). *Vietnam Electricity Annual Report 2016*. http://www.evn.com.vn/userfile/files/2017/3/AnnualReport2016.pdf

Ey, M. and Sherval, M. (2016). Exploring the minescape: Engaging with the complexity of the extractive Sector. *Area*, 48(2): 176–182.

Fairbairn, M. (2014). 'Like gold with yield': Evolving intersections between farmland and finance. *Journal of Peasant Studies*, 41(5): 777–795.

Fairhead, J. and Leach, M. (1996). *Misreading the African landscape: Society and ecology in a forest-savanna mosaic*. Cambridge: Cambridge University Press.

Fang, H., *et al.* (2015). Demystifying the Chinese housing boom. Working Paper 21112. Cambridge: National Bureau of Economic Research.

Febvre, L. (1973). Frontière: The world and the concept. In: Burke P. ed., *A new kind of histroy: From the writings of Lucie Febvre*. New York: Harper and Row, pp. 208–218.

Ferguson, J. (1994). *The anti-politics machine: Development, depoliticization, and bureaucratic power in Lesotho*. Minneapolis: University of Minnesota Press.

Ferguson, W. (1998). *Hokkaido highway blues: Hitchhiking Japan*. New York: Soho Press.

Ferguson, J. and Gupta, A. (2002). Spatializing states: Toward an ethnography of neoliberal governmentality. *American Ethnologist*, 29(4): 981–1002.

Fischer, M. (2003). *Emergent forms of life and the anthropological voice*. Durham, NC: Duke University Press.

Fishbein, G. (2010). *Berau forest carbon program: Delivering practical solutions to support development of a national-level REDD framework in Indonesia*. Jakarta: The Nature Conservancy and Republic of Indonesia Ministry of Forestry.

Fishbein, G. and Lee, D. (2015). *Early lessons from jurisdictional REDD+ and low emissions development programs*. Arlington, VA: The Nature Conservancy (TNC).

Fold, N. and Hirsch, P. (2009). Re-thinking frontiers in Southeast Asia: Editorial. *The Geographical Journal*, 175(2): 95–97.

Foucault, M. (1986). Of other spaces. *Diacritics*, 16: 22–27.

Foucault, M. (2008). *The birth of biopolitics: Lectures at the Collège de France 1978–1979*. New York: Picador.

Fox, J. and Castella, J.-C. (2013). Expansion of rubber (Hevea brasiliensis) in mainland Southeast Asia: What are the prospects for smallholders? *Journal of Peasant Studies*, 40(1): 155–170.

Fox, J., *et. al.* (2014). "Swidden, rubber and carbon: Can REDD+ work for people and the environment in Montane Mainland Southeast Asia?" *Global Environmental Change*, 29: 318–326.

Fujita, F. (1994). *American pioneers and the Japanese frontier: American experts in nineteenth-century Japan*. Westport, CT: Greenwood Press.

Fukuwaka, M. and Morita, K. (2008). Increase in maturation size after the closure of a high seas gillnet fishery on hatchery-reared chum salmon Oncorhynchus keta. *Evolutionary Applications*, 1: 376–387.

Geiger, D. (2008). *Frontier encounters: Indigenous communities and settlers in Asia and Latin America*. Copenhagen: IWGIA.

Geiger, H., *et al.* (2002). *Status of salmon stocks and fisheries in the North Pacific Ocean*. North Pacific Anadromous Fish Commission Technical Report No. 4.

Gibbs, H. K., *et al.* (2010). Tropical forests were the primary sources of new agricultural land in the 1980s and 1990s. *Proceedings of the National Academy of Sciences*, 107: 16732–16737.

Goldman, M. (2001). *Constructing an environmental state: Eco-governmentality and other transnational practices of a 'green' World Bank. Social Problems*, 48(4): 499–523.

Goldman, M., Nadasdy, P. and Turner, M., eds. (2011). *Knowing nature: Conversations at the intersection of political ecology and science studies*. Chicago, IL: University of Chicago Press.

Golley, J. (2016). China's environmental challenges: Under the dome with no way out? *The Asia-Pacific Journal*, 14(22): 3.

Gordillo, G. (2014). *Rubble: The afterlife of destruction*. Durham, NC: Duke University Press.

Goss, A. (2014). Building the world's supply of quinine: Dutch colonialism and the origins of a global pharmaceutical industry. *Endeavor*, 38(1): 8–18.

Graham, S. (2010). *Disrupted cities: When infrastructure fails*. London: Routledge.

Greenpeace. (2012). *Thirsty coal: A water crisis exacerbated by China's new mega coal power bases*. Amsterdam: Greenpeace International.

Guhathakurta, M. (2008). Globalization, class and gender relations: The shrimp industry in southwestern Bangladesh. *Development*, 51(2): 212–219.

Guhathakurta, M. (2011). The gendered nature of migration in southwestern Bangladesh: Lessons for a climate change policy. Paper presented at the Rethinking Migration: Climate, Resource Conflicts and Migration in Europe, 13–14 October, Berlin, Germany.

Gunarso, P., *et al.* (2013). *Oil palm and land use change in Indonesia, Malaysia and Papua New Guinea*. Singapore: Round Table on Sustainable Palm Oil.

Günel, G. (2012). A dark art: Field notes on carbon capture and storage negotiations at COP 17, Durban. *Ephemera*, 12(1): 33–41.

Günel, G. (2016). What is carbon dioxide? When is carbon dioxide?. *PoLAR*, 39(1): 33–45.

Günel, G. (2019). *Spaceship in the desert: Energy, climate change and urban design in Abu Dhabi*. Durham, NC: Duke University Press.

Guyot-Réchard, B. (2016). *Shadow states: India, China and the Himalayas, 1910–1962*. Cambridge: Cambridge University Press.

Hahm, H. H. and Kang, K. P. (2007). Fishermen, environmentalists, and government's perceptions of the sea: Focusing on the conflicts of Saemangeum Project. *Environmental Sociology ECO*, 11(2): 247–284.

Halim, S. (2004). Marginalization or empowerment? Women's involvement in shrimp cultivation and shrimp processing plants in Bangladesh. In: Hossain, K. T., Imam, M. H. and Habib, S. E. eds., *Women, gender and discrimination*. Rajshani: University of Rajshahi, pp. 95–112.

Hall, D. (2011). Land Grabs, land control, and Southeast Asian crop booms. *Journal of Peasant Studies*, 38(4): 837–857.

Hall, D. (2013). *Land*. Cambridge: Polity Press.

Hall, D., Hirsch, P. and Li, T. (2011). *Powers of exclusion: Land dilemmas in Southeast Asia*. Singapore: NUS Press.

Hallegatte, S., *et al.* (2012). Investment decision making under deep uncertainty: Application to climate change. *Policy Research Working Paper Series*. Washington, DC: The World Bank.

Hansen, T. B. and Stepputat, F. (2001). Introduction: States of imagination. In: Hansen, T. B. and Stepputat, F. eds., *States of imagination: Ethnographic explorations of the postcolonial state*. Durham, NC: Duke University Press. pp. 1–38.

Haraway, D. (2003). *The companion species manifesto: Dogs, people, and significant otherness*. Chicago, IL: Prickly Paradigm Press.

Haraway, D. (2016). *Staying with the trouble: Making kin in the Cthulucene*. Durham, NC: Duke University Press.

Harms, E. and Hussain, S. (2014). Introduction: Remote and edgy: New takes on old anthropological themes. *Hau: Journal of Ethnographic Theory*, 4(1): 361–367.

Harvey, D. (2001). Globalization and the 'spatial fix'. *Geographical Review*, 3: 23–30.

Harvey, D. (2005). *The new imperialism*. Oxford: Oxford University Press.

Harvey, P. and Knox, H. (2015). *Roads: An anthropology of infrastructure and expertise*. Ithaca, NY: Cornell University Press.

Ha Viet, Q. (2016). Brokering power in Vietnam's Northwest: The case of ethnic Tai cadres. PhD Thesis, The Australian National University.

Headrick, D. (1981). *The tools of empire: Technology and European imperialism in the nineteenth century*. Oxford: Oxford University Press.

Hedge, S. (2013). *Extrajudicial killings in Manipur*. New Delhi: Ministry of Home Affairs/Government of India.

Heidegger, M. (1977). *The question concerning technology and other essays*. New York: Harper & Row.

Herscher, A. and Siddiqi, A. (2014). Spatial violence. *Architectural Theory Review*, 19(3): 269–277.

Heynan, N., *et al.* (2007). *Neoliberal environments: False promises and unnatural consequences*. New York: Routledge.

Hirsch, P. (2009). Revisiting frontiers as transitional spaces in Thailand. *Geographical Journal*, 175(2): 124–132.

Hoffman, L. (2009). Governmental rationalities of environmental city-building in contemporary China. In: Jeffreys, E. ed., *China's Governmentalities: Governing change, changing government*. New York: Routledge, pp. 107–124.

Honigsbaum, M. (2002). *The fever trail: In search of the cure for malaria*. London: Pan Books.

Hossain, N. (2017). *The aid lab: Understanding Bangladesh's unexpected success*. Oxford: Oxford University Press.

Hostetter, E. (2011). Boomtown landscapes. *Journal of Material Culture*, 43(2): 59–79.

Howell, D. (1995). *Capitalism from within: Economy, society, and the state in a Japanese fishery*. Berkeley: University of California Press.

Hsing, Y. T. (2010). *The great urban transformation: Politics of land and property in China*. New York: Oxford University Press.

Huang, S., ed. (2010). *China coal outlook 2010*. Beijing: China Coal Industry Publishing House.

Human Rights Watch. (2008). *These fellows must be eliminated: Relentless violence and impunity in Manipur*. (No. 1-56432-379-X). New York: Human Rights Watch.

Hussain, S. (2015). *Remoteness and modernity: Transformation and continuity in Northern Pakistan*. New Haven, CT: Yale University Press.

Imphal Municipal Council. (2007). *City development plan: Imphal*. Imphal: Government of Manipur.

Institute for Defense Studies and Analyses IDSA. (2014). *Militant groups in South Asia*. New Delhi: IDSA.

Jacobs, J. M. (2006). A geography of big things. *Cultural Geographies*, 13(1): 1–27.

Jessop, B. (1990). *State theory: Putting capitalist states in their place*. Oxford: Polity Press.

Jilangamba, Y. (2010). Territorialities and identities: North-Eastern frontier of the British Indian Empire. PhD diss., New Delhi: Centre for Historical Studies, Jawaharlal Nehru University.

Jorgensen, A. and Rice, J. (2007). Uneven ecological exchange and consumption-based environmental impacts: A cross-national investigation. In: Hornborg, A., McNeill, J. R. and Martinez-Alier, J. eds., *Rethinking environmental history: World-system history and global environmental change*. Lanham, MD: Altamira Press, pp. 273–288.

Joss, S. (2015). *Sustainable cities: Governing for urban innovation*. London: Palgrave Macmillan.

Jullien, F. (1995). *The propensity of things: Toward a history of efficacy in China*. New York: Zone Books.

Kant, I. (1953 [1783]). *Prolegomena: To any future metaphysics that will be able to present itself as a science*. Manchester: Manchester University Press.

Karlsson, B. (2011). *Unruly hills: A political ecology of India's Northeast*. New York: Berghahn Books.

Karlsson, B. (2017). *A different story of coal: The power of power in Northeast India*. In: Neilson, K. and Oskarsson, P. eds., *Industrializing rural India: Land, policy, and resistance*. Abingdon: Routledge, pp. 107–122.

Kasimov, M. (1989). *Fruit conveyor of the golden valley [Fruktoviy Konveier Zolotoy Dolini]*. Dushanbe: Irfon.

Kassymbekova, B. (2011). Humans as territory: Forced resettlement and the making of Soviet Tajikistan, 1920–38. *Central Asian Survey*, 30(3–4): 349–370.

Kelly, A. and Peluso, N. (2015). Frontiers of commodification: State lands and their formalization. *Society & Natural Resources*, 28(5): 473–495.

Keough, S. B. (2015). Planning for growth in a natural resource boomtown: Challenges for urban planners in Fort McMurray, Alberta. *Urban Geography*, 36(8): 1169–1196.

Kikon, D. (2009). The predicament of justice: Fifty years of Armed Forces Special Powers Act in India. *Contemporary South Asia*, 17(3): 271–282.

Kincaid, J. (1988). *A small place*. New York: Farrar, Straus and Giroux.

King, G. (1876). *A manual of cinchona cultivation in India*. Calcutta: Office of Superintendent of Government Printing.

Klein, K. (1999). *Frontiers of historical imagination: Narrating the European conquest of Native America, 1890–1990*. Berkeley: University of California Press.

Koch, N. (2010). The monumental and the miniature: Imagining 'modernity' in Astana. *Social and Cultural Geography*, 11(8): 769–787.

Kocherga, F. K. (1966). Research work on the fight with erosion of soils and flooding in the mountainous regions of Central Asia. In: *Proceedings of the*

Central Asian Research Institute of Forestry Volume X. Tashkent: Uzbekistan Publishers, pp. 16–57.

Koh, L. P., *et al.* (2011). Remotely sensed evidence of tropical peatland conversion to oil palm. *Proceedings of the National Academy of Sciences of the United States of America*, 108(12): 5127–5132.

Kopytoff, I. (1999). The internal African frontier: Cultural conservatism and ethnic innovation. In: Rösler, M and Wendl, T. eds., *Frontiers and borderlands: Anthropological perspectives*. Frankfurt am Main: Peter Lang, pp. 31–44.

Korf, B. and Raeymaekers, T. (2013). Border, frontier and geography of rule at the margins of the state. In: Korf, B. and Raeymaekers, T. eds., *Violence on the margins: States, conflict, and borderlands*. New York: Palgrave Macmillan, pp. 3–28.

Krishna, S. (1996). Cartographic anxiety: Mapping the body politic in India. In: Alker, H. and Shapiro, M. eds., *Challenging boundaries: Global flows, territorial identities*. Minneapolis: University of Minnesota Press, pp. 193–214.

Lagerqvist, Y. (2013). Imagining the borderlands: Contending stories of a resource frontier in Muang Sing. *Singapore Journal of Tropical Geography*, 34(1): 57–69.

Lai, L., *et al.* (2014). The informational dimension of real estate development: A case of a 'positive non-interventionist' application of the Coase Theorem. *Land Use Policy*, 41: 225–232.

Latour, B. (1986). The powers of association. In: Law, J. ed., *Power, action and belief: A new sociology of knowledge?* London: Routledge, pp. 264–280.

Latour B. (2007). *Reassembling the social: An introduction to actor-network-theory*. New York: Oxford University Press.

Lattimore, O. (1940). *Inner Asian frontiers of China*. New York: American Geographical Society.

Laungaramsri, P. (2012). Frontier capitalism and the expansion of rubber plantations in Southern Laos. *Journal of Southeast Asian Studies*, 43(03): 463–477.

Law, J. (1994). *Organizing modernity*. Oxford: Blackwell.

Law, J. (2004). And if the global were small and non-coherent? Method, complexity and the baroque. *Environment and Planning D: Society and Space*, 22(1): 13–26.

Le Failler, P. (2011). The Đèo Family of Lai Châu: Traditional power and unconventional practices. *Journal of Vietnamese Studies*, 6(2): 42–67.

Le Failler, P. (2014). *La rivière noire: L'intégration d'une marche frontière au Vietnam*. Paris: CNRS Éditions.

Lee, M. (2015). *Dreams of the Hmong kingdom: The quest for legitimation in French Indochina, 1850–1910*. Madison: University of Wisconsin Press.

Lefebvre, H. (1978). *The production of space*. Oxford: Blackwell.

Lempert, R., *et al.* (2013). *Ensuring robust flood risk management in Ho Chi Minh City*. Policy Research Working Paper Series. Washington, DC: The World Bank.

Lentz, C. (2011a). Making the Northwest Vietnamese. *Journal of Vietnamese Studies*, 6(2): 68–105.

Lentz, C. (2011b). Mobilization and state formation on a frontier of Vietnam. *Journal of Peasant Studies*, 38(3): 559–586.

Lentz, C. (2014). The king yields to the village? A micropolitics of statemaking in Northwest Vietnam. *Political Geography*, 39: 1–10.

Lentz, C. (2017). Cultivating subjects: Opium and rule in post-colonial Vietnam. *Modern Asian Studies*, 51(4): 879–918.

Lentz, C. (2019). *Contested territory: Điện Biên Phủ and the making of Northwest Vietnam*. New Haven, CT: Yale University Press.

Levien, M. (2011). Special economic zones and accumulation by dispossession in India. *Journal of Agrarian Change*, 11(4): 454–483.

Levien, M. (2012). The land question: Special economic zones and the political economy of dispossession in India. *Journal of Peasant Studies*, 39(3–4): 933–969.

Li, R. (2013). Rational geographical distribution and efficient use of resources. In: Yue, F. and Cui, T. eds., *Blue book of coal industry*. Beijing: Social Sciences Academic Press, pp. 105–116.

Li, T. (2007a) Practices of assemblage and community forest management. *Economy and Society*, 36(2): 263–293.

Li, T. (2007b). *The will to improve: Governmentality, development, and the practice of politics*. Durham, NC: Duke University Press.

Li, T. (2014a). *Land's end: Capitalist relations on an indigenous frontier*. Durham, NC: Duke University Press.

Li, T. (2014b). What is land? Assembling a resource for global investment. *Transactions of the Institute of British Geographers*, 39(4): 589–602.

Li, T. (2015). Transnational farmland investment: A risky business. *Journal of Agrarian Change*, 15(4): 560–568.

Liang, H., *et al.* (2015). Effect of reclamation on hydrodynamics in the Shuanglong River Estuary. In *The 36th IAHR World Congress*, 8. The Hague

Lichatowich, J. (1999). *Salmon without rivers: a history of the Pacific salmon crisis*. Washington, DC: Island Press.

Lin, G. C. S. and Yi, F. (2011). Urbanization of capital or capitalization on urban land? Land development and local public finance in urbanizing China. *Urban Geography*, 32(1): 50–79.

Ludden, D. (2011). The process of empire: Frontiers and borderlands. In: Bang, P. and Bayly, C. eds., *Tributary empires in global history*. Cambridge: Cambridge University Press, pp. 132–150.

Luke, T. (1995). On environmentality: Geo-power and eco-knowledge in the discourses of contemporary environmentalism. *Cultural Critique*, (31): 57–81.

Lund, C. (2006). Twilight institutions: Public authority and local politics in Africa. *Development and Change*, (37)4: 685–705.

Lund, C. (2011). Fragmented sovereignty: Land reform and dispossession in Laos. *Journal of Peasant Studies*, 38(4): 885–905.

Lund, C. (2014). Of what is this a case? Analytical movements in qualitative social science research. *Human Organization*, 73(3): 224–234.

Lunminthang, M. (2016). Rethinking the political history of Northeast India: Historical review on Kuki Country. *Indian Historical Review*, 43(1): 63–82.

MacKinnon, J., Verkuil, Y. I., and Murray, N. (2012). IUCN situation analysis on East and Southeast Asian intertidal habitats, with particular reference to the Yellow Sea (including the Bohai Sea). *Occasional Paper of the IUCN Species Survival Commission 47*.

MacLean, K. (2013). *The government of mistrust: Illegibility and bureaucratic power in socialist Vietnam*. Madison: University of Wisconsin Press.

Magee, D. (2006). Powershed politics: Yunnan hydropower under great western development. *The China Quarterly*, 185: 23–41.

Magenda, B. D. (1991). *East Kalimantan: The decline of a commercial aristocracy*. Sheffield: Equinox Publishing.

Maki, J. M. (2002). *A yankee in Hokkaido: The life of William Smith Clark*. Oxford: Lexington Books.

Marcus, G. and Saka, E. (2006). Assemblage. *Theory, Culture, Society*, 23(2–3): 101–109.

Margono, B. A., *et al.* (2014). Primary forest cover loss in Indonesia over 2000–2012. *Nature Climate Change*, 4: 730–735.

Markham, C. (1880). *Peruvian bark: A popular account of the introduction of Chinchona cultivation into British India*. London: John Murray.

Marshall, A. J. (2002). *Summary of orangutan surveys conducted in Berau District, East Kalimantan November 2001–September 2002*. The Nature Conservancy (TNC).

Marshall, A. J., *et al.* (2007). Use of limestone karst forests by Bornean orangutans (Pongo pygmaeus morio) in the Sangkulirang Peninsula, East Kalimantan, Indonesia. *American Journal of Primatology*, 69(2): 212–219.

Martin, R. (2007). *An empire of indifference: American war and the financial logic of risk management*. Durham, NC: Duke University Press.

Marx K. (1976) *Capital: A critique of political economy*. London: Penguin.

Mason, M. (2012). *Dominant narratives of colonial Hokkaido and imperial Japan: Envisioning the periphery and the modern nation-state*. London: Palgrave Macmillan.

Mason, M. and Lee, H., eds. (2012). *Reading colonial Japan: Text, context, and critique*. Stanford, CA: Stanford University Press.

Massey D. (1994). *Space, place, and gender*. Minneapolis: University of Minnesota Press.

Massey, D. (2007). *World city*. Cambridge: Polity Press.

Massumi, B. (2009). National enterprise emergency: Steps toward an ecology of powers. *Theory, Culture, Society*, 26: 153–186.

Mathews, A. S. and Barnes, J. (2016). Prognosis: Visions of environmental futures. *Journal of the Royal Anthropological Institute*, 22(S1): 9–26.

Mathur, N. (2016). *Paper tiger: Law, bureaucracy and the developmental state in Himalayan India*. New Delhi: Cambridge University Press.

Mathur, S. (2012). Life and death in the borderlands: Indian sovereignty and military impunity. *Race & Class*, 54(1): 33–49.

Mattei, U. and Nader, L. (2008). *Plunder: When the rule of law is illegal*. London: Blackwell.

McCarthy, J. (2006). *The fourth circle: A political ecology of Sumatra's rainforest frontier*. Stanford, CA: Stanford University Press.

McCarthy, J. and Cramb, R. (2009). Policy narratives, landholder engagement, and oil palm expansion on the Malaysian and Indonesian Frontiers. *Geographical Journal*, 175(2): 112–123.

McCarthy, J. F., Vel, J. A. and Afiff, S. (2012). Trajectories of land acquisition and enclosure: Development schemes, virtual land grabs, and green acquisitions in Indonesia's Outer Islands. *Journal of Peasant Studies*, 39(2): 521–549.

McDuie-Ra, D. (2009). Fifty-year disturbance: The Armed Forces Special Powers Act and exceptionalism in a South Asian periphery. *Contemporary South Asia*, 17(3): 255–270.

McDuie-Ra, D. (2012). *Northeast migrants in Delhi: Race, refuge and retail*. Amsterdam: Amsterdam University Press.

McDuie-Ra, D. (2014). Borders, territory, and ethnicity: Women and the Naga peace process. In: Naples, N. and Mendez, J. B. eds., *Border politics: Social movements, collective identities, and globalization*. New York: New York University Press, pp. 95–119.

McDuie-Ra, D. (2016). *Borderland city in New India: Frontier to gateway*, Amsterdam: Amsterdam University Press.

McDuie-Ra, D. and Kikon, D. (2016). Tribal communities and coal mining in Northeast India: The politics of imposing and resisting mining bans. *Energy Policy*, 99: 261–269.

McElwee, P. (2004). You say illegal, I say legal: The relationship between 'illegal' logging and poverty, land tenure, and forest use rights in Vietnam. *Journal of Sustainable Forestry*, 19(1/2/3): 97–135.

McElwee, P. (2016). *Forests are gold: Trees, people, and environmental rule in Vietnam*. Seattle: University of Washington Press.

McEnaney, L. (2000). *Civil defence begins at home: Militarization meets everyday life in the 1950s*. Princeton, NJ: Princeton University Press.

McGranahan, C. (2010). *Arrested histories: Tibet, the CIA, and memories of a forgotten war*. Durham, NC: Duke University Press.

McGregor, A., Eilenberg, M. and Coutinho, J. B., eds. (2015). From global policy to local politics: The social dynamics of REDD+ in Asia Pacific [Special Issue]. *Asia Pacific Viewpoint*, 56(1): 1–152.

Middleton, T. (2015). *Demands of recognition: State anthropology and ethnopolitics in Darjeeling*. Stanford, CA: Stanford University Press.

Mitchell, D. (2008). New axioms for reading the landscape: Paying attention to political economy and social justice. In: Wescoat, J. L. Jr. and Johnston, D. M. eds., *Political economies of landscape change*. Dordrecht: Springer, pp. 29–50.

Mitchell, T. (1991). The limits of the state: Beyond statist approaches and their critics. *American Political Science Review*, 85(1): 77–96.

Mitchell, T. (2002). *Rule of experts: Egypt, techno-politics, modernity*. Berkeley: University of California Press.

Mitchell, T. (2011). *Carbon democracy: Political power in the age of oil*. London: Verso.

Moniaga, S. (1993). Towards community-based forestry and recognition of Adat property rights in the outer islands of Indonesia. In: Fox, J. ed., *Legal frameworks for forest management in Asia: Case studies of community/ state relations*. Honolulu: Program on Environment Occasional Paper no. 16, Honolulu: East-West Center, pp. 131–150.

Moore, D. (2005). *Suffering for territory: Race, place, and power in Zimbabwe*. Durham, NC: Duke University Press.

Moore, J. W. (2000). Sugar and the expansion of the early modern world-economy: Commodity frontiers, ecological transformation, and industrialization. *Review (Fernand Braudel Center)*, 23(3): 409–433.

Morris-Suzuki, T. (1994). Creating the frontier: Border, identity and history in Japan's far north. *East Asian History*, 7: 1.

Morris-Suzuki, T. (1996). A descent into the past: The Frontier in the construction of Japanese identity. In: Denoon, D., *et al.* eds., *Multicultural Japan: Palaeolithic to postmodern*. Melbourne: Cambridge University Press.

Morris-Suzuki, T. (1998). *Re-inventing Japan: Time, space, nation*. New York: M.E. Sharpe.

Morris-Suzuki, T. (1999). Lines in the snow: Imagining the Russo-Japanese Frontier. *Pacific Affairs*, 72(1): 57–77.

Mosse, D. (2004). Is good policy unimplementable? Reflections on the ethnography of aid policy and practice. *Development and Change*, 35(4): 639–671.

Mukherjee, A. (1998). The Peruvian Bark revisited: A critique of British cinchona policy in colonial India. *Bengal Past and Present*, 117(1–2): 81–102.

Mukherjee (Mukhopadhyay), A. (1996). Natural science in colonial context: The Calcutta Botanic Garden and Agri-Horticultural Society of India, 1878–1870. PhD diss., Jadavpur University.

Müller, M. (2015). Assemblages and actor-networks: Rethinking socio-material power, politics and space. *Geography Compass*, 9(1): 27–41.

Müller, M. and Schurr, C. (2016). Assemblage thinking and actor-network theory: Conjunctions, disjunctions, cross-fertilisations. *Transactions of the Institute of British Geographers*, 41(3): 217–229.

Nail, T. (2017). What is an assemblage? *SubStance*, 46(1): 21–37.

National Rural Health Mission NRHM. (2012). *Regional Evaluation Team Report 2011–12*. New Delhi: NRHM.

Navratil, P., *et al.* (2012). *Survey on the land cover situation and land-use change in the district Berau, Indonesia*. Hamburg: GIZ FORCLIME.

Nazpary, J. (2002). *Post-Soviet chaos: Violence and dispossession in Kazakhstan*. London: Pluto Press.

Neumann, R. P. (1998). *Imposing wilderness: Struggles over livelihood and nature preservation in Africa*. Berkeley: University of California Press.

Nigmatov, U. (1977). Results and prospects of research on Department of Forestry at SredAzHIILh. In: *Proceedings of the Central Asian Research Institute of Forestry Volume X*. Tashkent: Uzbekistan Publishers, pp. 128–143.

Noakes, D., *et al.* (1999). A preliminary look at Pacific salmon catches in 1999. North Pacific Anadromous Fish Commission Doc. 452 Rev. 1.

Norwegian Refugee Council. (2015). *Community resilience and disaster-related displacement in South Asia*. Oslo: Norwegian Refugee Council.

O'Brien, K. J. (2008). *Popular protest in China*. Cambridge, MA: Harvard University Press.

OECD. (2011). *Towards green growth*. Paris: OECD.

Ogden, L. A. (2011). *Swamplife: People, gators, and mangroves entangled in the everglades*. Minneapolis: University of Minnesota Press.

Ong, A. (2006). *Neoliberalism as exception: Mutations in citizenship and sovereignty*. Durham, NC: Duke University Press.

Ong, A. (2011). Hyperbuilding: Spectacle, speculation, and the hyperspace of sovereignty. In A. Roy and A. Ong eds., pp. 205–226. *Worlding cities: Asian Experiments and the art of being global*. Malden, MA: Wiley-Blackwell.

Ong, A. and Collier S. J. (2005). *Global assemblages: Technology, politics, and ethics as anthropological problems*. Malden, MA: Blackwell.

Pachuau, J. and Van Schendel, W. (2015). *The camera as witness: A social history of Mizoram, Northeast India*. Delhi: Cambridge University Press.

Paprocki, K. (2017). Threatening dystopias: Development politics and the anticipation of climate crisis in Bangladesh. PhD diss., Cornell University.

Paprocki, K. (2018). Threatening dystopias: Development and adaptation regimes in Bangladesh. *Annals of the Association of American Geographers*, 108(4): 955–973.

Paprocki, K. (2019). All that is solid melts into the bay: Anticipatory ruination and climate change adaptation. *Antipode*, 51(1).

Paprocki, K. and Cons, J. (2014a). Life in a shrimp zone: Aqua- and other cultures of Bangladesh's coastal landscape. *Journal of Peasant Studies*, 41(6): 1109–1130.

Paprocki, K. and Cons, J. (2014b). Brackish waters and salted lands: The social cost of shrimp in Bangladesh. *Food First Backgrounder*. Food First. Vol. 20.

Paprocki, K. and Huq, S. (2017). Shrimp and coastal adaptation: On the politics of climate justice. *Climate and Development*, 10(1): 1–3.

Park, S. Y. (2007). Structure and peculiarities of social conflict around Saemanguem Reclamation Project TT – structure and peculiarities of social conflict around Saemanguem Reclamation Project. *Environmental Sociology ECO*, 11(1): 133–166.

Parratt, S. N. A. (1980). *The religion of Manipur: Beliefs, rituals, and historical Development*. Kolkata: Firma KLM.

Patel, R. and Moore, J. (2017). *A history of the world in seven cheap things: A guide to capitalism, nature, and the future of the planet*. Berkeley: University of California Press.

Peet, R. and Watts, M., eds. (1996). *Liberation ecologies: Environment, development, social movements*. New York: Routledge.

Pelley, P. (2002). *Postcolonial Vietnam: New histories of the national past*. Durham, NC: Duke University Press.

Peluso, N. (1992). *Rich forests, poor people*. Berkeley: University of California Press.

Peluso, N. (1996). Fruit trees and family trees in an anthropogenic forest: Ethics of access, property zones, and environmental change in Indonesia. *Comparative Studies in Society and History*, 38(3): 510–548.

Peluso, N. (2018). Entangled territories in small-scale gold mining frontiers: Labor practices, property, and secrets in Indonesian gold country. *World Development*, 101: 400–416.

Peluso, N. and Vandergeest, P. (2011). Political ecologies of war and forests: Counterinsurgencies and the making of national natures. *Annals of the Association of American Geographers*, 101(3): 587–608.

Peluso, N. and Watts, M., eds. (2001). *Violent environments*. Ithaca, NY: Cornell University Press.

Perreault, T., Bridge, G. and McCarthy, J., eds. (2015). *The Routledge handbook of political ecology*. New York: Routledge.

Phanjoubam, P. (2005). Manipur: Fractured land. *India International Centre Quarterly*, 32(2–3): 275–287.

Potter, L. (1988). Indigenes and Colonizers: Dutch forest policy in South and East Borneo (Kalimantan) 1900 to 1950. In: Dargavel, J., Dixon, K. and Semple, N. eds., *Changing tropical forests: Historical perspectives on today's challenges in Asia, Australia, and Oceania*. Canberra: The Australian National Unversity.

Prescott, J. (1987). *Political frontiers and boundaries*, London: Allen and Unwin.

Province of East Kalimantan. (2011). Kaltim Hijau. Samarinda, Indonesia: government of East Kalimantan. http://www.kaltimprov.go.id/halaman-20-kaltim-green.html

Prudham, S. (2003). Taming trees: Capital, science, and nature in Pacific slope tree improvement. *Annals of the Association of American Geographers*, 93(3): 636–656.

Pye, O. and Bhattacharya, J. (2012). *The palm oil controversy in Southeast Asia: A transnational perspective*. Singapore: ISEAS Publishing.

Quan, H. (2016). Brokering power in Vietnam's northwest: The case of ethnic Tai cadres. PhD diss., The Australian National University.

Rajan, K. S. (2012). *Lively capital: Biotechnologies, ethics, and governance in global markets*. Durham, NC: Duke University Press.

Ranganathan, M. (2015). Storm drains as assemblages: The political ecology of flood risk in post-colonial Bangalore. *Antipode*, 47(5): 1300–1320.

Rasmussen, M. B. and Lund, C. (2018). Reconfiguring frontier spaces: The territorialization of resource control. *World Development*, 101: 388–399.

Redclift, M. (2006). *Frontiers: Histories of civil society and nature*. Cambridge, MA: MIT Press.

Reeves, M. (2014). *Border work: Spatial lives of the state in rural Central Asia*. Ithaca, NY: Cornell University Press.

Ribot, J. (2014). Cause and response: Vulnerability and climate in the Anthropocene. *Journal of Peasant Studies*, 41: 667–705.

Richards, J. F. (2003). *The unending frontier: An environmental history of the early modern world*. Berkeley: University of California Press.

Rieber, A. (2001). Frontiers in history. In: Baltes, P. B. ed., *International encyclopedia of the social and behavioral sciences*. Oxford: Pergamon, pp. 5812–5818.

Riles, A. (2000). *The network inside out*. Ann Arbor: University of Michigan Press.

Robertson, M. M. (2012). Measurement and alienation. Making a world of ecosystem services. *Transactions of the Institute of British Geographers*, 37(3): 386–401.

Rocco, F. (2004). *The miraculous fever-tree: Malaria, medicine and the cure that changed the world*. London: Harper Collins.

Rose, N. (1999). *Powers of freedom: Reframing political thought*. Cambridge: Cambridge University Press.

Rossabi, M., ed. (2004). *Governing China's multiethnic frontiers*. Seattle: University of Washington Press.

Rowe, W. C. (2009). Kitchen gardens in Tajikistan: The economic and cultural importance of small-scale private property in a post-Soviet society. *Human Ecology*, 37(6): 691–703.

Roy, A. (2011). The blockade of the world-class city: Dialectical images of Indian urbanism. In: Roy, A. and Ong, A. eds., *Worlding cities: Asian experiments and the art of being global*. Malden: Wiley Blackwell, pp. 259–278.

Roy A. (2012). Subjects of risk: Technologies of gender in the making of millennial modernity. *Public Culture*, 24(1): 131–155.

Roy, A. and Ong, A., eds. (2011). *Worlding cities: Asian experiments and the art of being global*. Malden, MA: Wiley Blackwell.

Roy, O. (2000). *The new central Asia: The creation of nations*. London: Tauris.

Rubinov, I. (2014). Migrant assemblages: Building postsocialist households with Kyrgyz remittances. *Anthropological Quarterly*, 87(1): 183–215.

Rubinov, I. (2016). *The impact of migration and remittances on natural resources in Tajikistan*. Occasional Paper no. 164. Bogor: Center for International Forestry Research (CIFOR).

Sayer, D. (1994). Everyday forms of state formation: Dissident remarks on hegemony. In: Joseph, G. M. and Nugent, D. eds., *Everyday forms of state formation: Revolution and the negotiation of rule in modern Mexico*. Durham, NC: Duke University Press, pp. 367–377.

Scott, J. (1998). *Seeing like a state: How certain schemes to improve the human condition have failed*. New Haven, CT: Yale University Press.

Scott J. (2009). *The art of not being governed: An anarchist history of upland Southeast Asia*, New Haven, CT and London: Yale University Press.

Seitz, W. C. (1961). *The art of assemblage*. New York: Museum of Modern Art.

Shanklin, E. (1988). Beautiful deadly Lake Nyos: The explosion and its aftermath. *Anthropology Today*, 4(1): 12–14.

Siddiqui, T. (2003). Migration as a livelihood strategy of the poor: The Bangladesh case. Paper presented at the Regional Conference on Migration, Development and Pro-Poor Policy Choices in Asia, 22–24 June, Dhaka.

Simpson, T. (2008). Macao, capital of the 21st century? *Environment and Planning D: Society and Space*, 26(6): 1053–1079.

Sivaramakrishnan, K. (1999). *Modern forests: Statemaking and environmental change in colonial Eastern India.* Stanford, CA: Stanford University Press.

Smil, V. (1999). China's agricultural land. *The China Quarterly,* 158: 414–429.

Smith W. and Dressler, W. H. (2017). Rooted in place? The coproduction of knowledge and space in agroforestry assemblages. *Annals of the American Association of Geographers,* 107(4): 897–914.

Spruce, R. (1861). *Report on the expedition to procure seeds and plants of the Cinchona succirubra, or red bark tree.* London: Eyre and Spottiswoode.

Stalcup, M. (2015). Policing uncertainty: On suspicious activity reporting. In: Samimian-Darash, L. and Rabinow, P. eds., *Modes of uncertainty: Anthropological cases.* Chicago, IL: University of Chicago Press, pp. 69–87.

Stein, S. (2015). Coping with the 'world's biggest dust bowl': Towards a history of China's forest shelterbelts, 1950s–Present. *Global Environment,* 8: 320–348.

Steinberg, P. E. (2001). *The social construction of the ocean.* Cambridge: Cambridge University Press.

Stewart, S. (1993). *On longing: Narratives of the miniature, the gigantic, the souvenir, the collection.* Durham, NC: Duke University Press.

Stoler, A. (2013). 'The rot remains': From ruins to ruination. In: Stoler, A. ed., *Imperial debris: On ruins and ruination.* Durham, NC: Duke University Press. pp. 1–35.

Stoler, A. (2016). *Duress: Imperial durabilities in our times.* Durham, NC: Duke University Press.

Stoler, A. and Cooper, F. (1997). *Tensions of empire: Colonial cultures in a bourgeois world.* Berkeley: University of California Press.

Strathern, M. (2013). Environments within: An ethnographic commentary on scale. In: Appendix III, *Learning to see in Melanesia.* HAU Masterclass Series.

Sturgeon J. C. (2005). *Border landscapes: The politics of Akha land use in China and Thailand.* Seattle: University of Washington Press.

Sur, M. (2010). Chronicles of repression and resilience. In: Hossain, H. Guhathakurta, M. and Sur, M. eds., *Freedom from fear, freedom from want? Re-thinking security in Bangladesh.* New Delhi: Rupa, pp. 47–61.

Suryanata, K. (1994). Fruit trees under contract: Tenure and land use change in upland Java, Indonesia. *World Development,* 22(10):1567–1578.

Swanson, H. A. (2015). Shadow ecologies of conservation: Co-production of salmon landscapes in Hokkaido, Japan and Southern Chile. *Geoforum,* 61:101–110.

Swanson, HA. (2018). Landscapes By Comparison: Practices of Enacting Salmon in Hokkaido, Japan. In Omura, K. et al., eds. The World Multiple: The Quotidian Politics of Knowing and Generating Entangled Worlds. New York: Routledge, pp. 105–122.

Swyngedouw, E. (1999). Modernity and hybridity: Nature, regeneracionismo, and the production of the Spanish waterscape. *Annals of the Association of American Geographers,* 89(3):443–465.

Sze, J. (2015). *Fantasy islands: Chinese dreams and ecological fears in an age of climate crisis.* Berkeley: University of California Press.

Tagliacozzo E. (2005). *Secret trades, porous borders: Smuggling and states along a Southeast Asian frontier, 1865–1915*, New Haven, CT: Yale University Press.

Tambiah, S. (2013). The galactic polity in Southeast Asia. *Hau: Journal of Ethnographic Theory*, 3(3): 503–534.

The Economist. (2015). Hydropower in Vietnam: Full to bursting. *The Economist* Jan.

Tilley, H. (2011). *Africa as a living laboratory: Empire, development, and the problem of scientific knowledge, 1970–1950*. Chicago, IL: University of Chicago Press.

TNC. (2015). *Our partners in conservation*. http://www.nature.org/about-us/our-partners/index.htm

Toland, A. (2017). Hong Kong's artificial anti-archipelago and the unnaturing of the natural. In: Rademacher, A. and Sivaramakrishnan, K. eds., *Places of nature in ecologies of urbanism*. Hong Kong: Hong Kong University Press, pp. 87–107.

Tsing, A. (1993). *In the realm of the Diamond Queen: Marginality in an out-of-the-way place*. Princeton, NJ: Princeton University Press.

Tsing, A. (1997). Transitions as translations. In: Scott, J., Kaplan, C. and Keates, D. eds., *Transitions, environments, translations: Feminisms in international politics*. New York: Routledge, pp. 253–272.

Tsing, A. (2000). Inside the economy of appearance. *Public Culture*, 12 (1): 115–144.

Tsing, A. (2003). Natural resource and capitalist frontiers. *Economic & Political Weekly*, 38(48): 5100–5106.

Tsing, A. (2005). *Friction: An ethnography of global connection*. Princeton, NJ: Princeton University Press.

Tsing, A. (2015). *The mushroom at the end of the world: On the possibility of life in capitalist ruins*. Stanford, CA: Princeton University Press.

Tu, J. (2011). Industrial organization of the Chinese coal industry. Working Paper 103. Program on Energy and Sustainable Development. Stanford: Freeman Spogli Institute for International Studies.

Turner F. J. (1920). *The frontier in American history*, New York: Holt.

UNDP/GEF Yellow Sea Project. (2007). *Reducing environmental stress in the Yellow Sea large marine ecosystem: Transboundary analysis*. Ansan: UNDP/GEF Yellow Sea Project.

United Nations Environment Programme UNEP. (2011). *Towards a green economy: Pathways to sustainable development and poverty eradication*. Nairobi: United Nations Environment Programme.

van der Hoogte, A. and Pieters, T. (2015). Science, industry and the colonial state: A shift from a German- to a Dutch-controlled Cinchona and Quinine cartel (1880–1920). *History of Technology*, 31(1): 2–36.

van Noordwijk, M., *et al*. (2013). Reducing emissions from land use in Indonesia: Motivation, policy instruments and expected funding streams. *Mitigation and Adaptation Strategies for Global Change*, 19(6): 677–692.

Veale, L. (2010). A historical geography of the Nilgiri cinchona plantations, 1860–1900. PhD diss., University of Nottingham.

Vellakkal, S. and Ebrahim, S. (2013). Publicly-financed health insurance schemes. *Economic & Political Weekly*, 48(1): 24–27.

Verbrugge B. (2015). Undermining the state? Informal mining and trajectories of state formation in Eastern Mindanao, Philippines. *Critical Asian Studies*, 47(2): 177–199.

Vihrestok, S. P. and Padalko, V. V. (1983). *Rational use, protection and development of forest resources in Tajikistan*. Dushanbe: TadjikNIINTI.

van Schendel, W. (1982). *Peasant mobility*. New Delhi: Manohar.

van Schendel, W. (2002). Geographies of knowing, geographies of ignorance: Jumping scale in Southeast Asia. *Environment and Planning D: Society and Space*, 20(6): 647–668.

von Schnitzler, A. (2016). *Democracy's infrastructure: Techno-politics and protest after apartheid*. Durham, NC: Duke University Press.

Walker, B. L. (2001). *The conquest of Ainu lands: Ecology and culture in Japanese expansion, 1590–1800*. Berkeley: University of California Press.

Wallerstein, I. (1974). *The modern world-system 1: Capitalist agriculture and the origins of the European world-economy in the sixteenth century*. New York: Academic Press.

Wang, T. (2009). Zhongguo Shamo yu Shamohua Kexue 50 Nian (Fifty years of desert and desertification science in China). In: Wang, T, ed., *Zhongguo hanqu Hanqu huanjing yu gongcheng eexue 50 Nian (Fifty years of cold and arid environmental and engineering science in China)*. Beijing: Kexu Chuban She (Scientific Press), pp. 28–60.

Wang, W., *et al.* (2014). Development and management of land reclamation in China. *Ocean & Coastal Management*, 102: 415–425.

Water, Land, and Ecosystems WLE. (2017). *Dataset on the dams of the Irrawaddy, Mekong, Red, and Salween River Ecosystems*. Vientiane: CGIAR Research Program on WLE.

Watts, M. (1983). On the poverty of theory: Natural hazards research in context. In: Hewitt, K. ed., *Interpretations of calamity: from the viewpoint of human ecology*. Boston, MA: Allen & Unwin, pp. 231–262.

Watts, M. (2014). Oil frontiers: The Niger Delta and the Gulf of Mexico. In: Barrett, R. and Worden, D. eds., *Oil culture*, 1st ed. Minneapolis: University of Minnesota Press, pp. 180–210.

Watts, M. (2015a). Specters of oil: An introduction to the photographs of Ed Kashi. In: Appel, H., Mason, A. and Watts, M. eds., *Subterranean estates: Life worlds of oil and gas*. Ithaca: Cornell University Press, pp. 165–188.

Watts, M. (2015b). Now and then: The origins of political ecology and the rebirth of adaptation as a form of thought. In: Perreault, T., Bridge, G. and McCarthy, J. eds., *The Routledge handbook of political ecology*. New York: Routledge, pp. 19–50.

Wendl, T. and Rösler, M. (1999). Introduction: Frontiers and borderlands: The rise and relevance of an anthropological research genre. In: Rösler, M. and Wendl, T. eds., *Frontiers and borderlands: Anthropological perspectives*. Frankfurt am Main: Peter Lang, pp. 1–29.

White, R. (1995). *The organic machine: The remaking of the Columbia River*. New York: Hill and Wang.

Wicke, B., *et al.* (2011). Exploring land use changes and the role of palm oil production in Indonesia and Malaysia. *Land Use Policy*, 28: 193–206.

Williams, R. (1983). *The country and the city*. New York: Oxford University Press.

Williams, J. G., *et al.* (2008). Potential for anthropogenic disturbances to influence evolutionary change in the life history of a threatened salmonid. *Evolutionary Applications*, 1: 271–285.

Wisner, B., *et al.* (2004). *At risk: Natural hazards, people's vulnerability and disasters*. London: Routledge.

Wolf, E. (1982). *Europe and the people without history*. Berkeley: University of California Press.

Wolf, E. (1999). *Envisioning power: Ideologies of dominance and crisis*. Berkeley: University of California Press.

Wolford, W., *et al.* (2013a). *Governing global land deals: The role of the state in the rush for land*. Oxford: Wiley Blackwell.

Wolford, W., *et al.* (2013b). Governing global land deals: the role of the state in the rush for land. *Development and Change*, 44(2): 189–210.

Woods, K. (2011a). Ceasefire capitalism: Military-private partnerships, resource concessions and military-state building in the Burma-China borderlands. *Journal of Peasant Studies*, 38(4): 747–770.

Woods, K. (2011b). Conflict timber along the China-Burma border. In: Tagliacozzo, E. and Wen-Chin, C. eds., *Chinese circulations: capital, commodities and networks in Southeast Asia*. Durham, NC: Duke University Press, pp. 480–507.

Woodworth, M. D. (2015). China's coal production goes west: Assessing recent geographical restructuring and industrial transformation. *The Professional Geographer*, 67(4): 630–640.

Woodworth, M. D. (2017). Disposable Ordos: The making of an energy resource frontier in Ordos, Inner Mongolia. *Geoforum*, 78: 133–140.

Xin, L. (2009). *The mirage of China: Anti-humanism, narcissism, and corporeality of the contemporary world*. New York: Berghahn Books.

Yaguchi, Y. (2000). Remembering a more layered past in Hokkaido: Americans, Japanese, and the Ainu. *The Japanese Journal of American Studies*, 11: 109–128.

Yeh, E. (2013). *Taming Tibet: Landscape transformation and the gift of Chinese development*. Ithaca, NY: Cornell University Press.

Yusupov, H. S. (1966). 25 years of the Central Asian Research Institute of Forestry. In: *Proceedings of the Central Asian Research Institute of Forestry Volume X*. Tashkent: Uzbekistan Publishers, pp. 3–15.

Zee, J. C. (2017). Holding patterns: Sand and political time at China's desert shores. *Cultural Anthropology*, 32(2): 215–241.

Zeiderman, A. (2016). *Endangered city: The politics of security and risk in Bogotá*. Durham, NC: Duke University Press.

Zhan, C. and de Jong, M. (2017). Financing Sino-Singapore Tianjin eco-city: What lessons can be drawn for other large-scale sustainable city-projects? *Sustainability*, 9(2): 201–217.

Zhang, C. X., *et al.* (2011). Steel plants in a circular economy society in China. *Iron and Steel/Gangtie*, 46(7): 1–6.

Zhang, Li. (2006). Contesting Spatilal Modernity in Late-Socialist China. Current Anthropology.

Zhu, G., *et al.* (2017). Assessment ecological risk of heavy metal caused by high-intensity land reclamation in Bohai Bay, China. *PLoS One*, 12(4): e0175627.

Zhu, X. and Sun, B. (2009). Tianjin Binhai new area: A case study of multi-level streams model of Chinese decision-making. *Journal of Chinese Political Science*, 14(2): 191–211.

Zimmerer, K. and Bassett, T. (2003). *Political ecology: An integrative approach to geography and environment.* New York: Guilford Press.

Zou, D. V. and Kumar, M. S. (2011). Mapping a colonial borderland: Objectifying the geo-body of India's Northeast. *The Journal of Asian Studies*, 70(1): 141–170.

Index

Frontier Assemblages: The Emergent Politics of Resource Frontiers in Asia,
First Edition. Edited by Jason Cons and Michael Eilenberg.
© 2019 John Wiley & Sons Ltd. Published 2019 by John Wiley & Sons Ltd.